John van Emden • Jennifer Easteal

Technical Writing and Speaking
An Introduction

The McGraw-Hill Companies

London • New York • St Louis • San Francisco • Auckland
Bogotá • Caracas • Lisbon • Madrid • Mexico • Milan
Montreal • New Delhi • Panama • Paris • San Juan
São Paulo • Singapore • Sydney • Tokyo • Toronto

Published by
McGraw-Hill Publishing Company
Shoppenhangers Road, Maidenhead, Berkshire, SL6 2QL, England
Telephone 01628 23432
Fax 01628 770224

British Library Cataloguing in Publication Data
Van Emden, Joan
   Technical writing and speaking: an introduction
   1. Technical writing 2. Technical correspondence 3. Public
speaking
   I. Title II. Easteal, Jennifer
   808'.066

   ISBN 0-07-709027-6

Library of Congress Cataloging-in-Publication Data
Van Emden, Joan.
   Technical writing and speaking: an introduction/Joan van Emden,
Jennifer Easteal.
     p. cm.
   Includes bibliographical references.
   ISBN 0-07-709027-6 (pbk. : alk. paper)
   1. Technical writing. 2. Communication of technical information.
3. Public speaking. I. Easteal, Jennifer. II. Title.
   T11.V363 1996
   808'.0666–dc20

                                 96-22751
                                        CIP

McGraw-Hill

A Division of The McGraw-Hill Companies

12345 CUP 99876
Typeset by BookEns Ltd, Royston, Herts
Printed and bound in Great Britain at the University Press, Cambridge

Printed on permanent paper in compliance with ISO Standard 9706

Contents

4
Written Text 50

5
Writing for Publication 82

6
Speaking Techniques 93

7
Spoken Presentation 123

8
Specimen Report and Presentation 145

Preface

Good communication results in the goodwill of the reader or the audience. It is all too easy, given modern technological resources, to overlook the importance of identifying the recipient and clarifying the objectives, choosing an appropriate form of communication and then using it in an efficient and effective way. In this book, we have discussed traditional means of transmitting information—letter, report or formal presentation—for many technical people use such methods regularly in their daily work. We have also considered more modern ways of communicating, such as fax, e-mail and videoconferencing. These have been around for too short a time to have developed their own conventions, and they are frequently used in an unsuitable or inefficient way.

Our final chapter brings together technical writing and speaking. A case study shows how an engineer prepares a report on his work and then gives a presentation, first to colleagues and then to his client company; stress is laid on the different stages of selecting, planning and preparing for these activities, so that readers are given a detailed understanding of the way in which the same material is adapted for very different forms of communication.

We have assumed three main groups of readers: those whose work involves them in day-to-day communication with colleagues, clients and customers; research students and lecturers, publishing or speaking at conferences for the first time, who may also be helping young undergraduates to communicate their growing technical knowledge; and students themselves, who often find it unexpectedly difficult to present their material in a suitable written or spoken form.

Our thanks are due to all the companies which have given us their material to use as examples; to colleagues at the University of Reading and especially to Dr A. J. Pretlove of the Department of Engineering and Mr Colin Gray of the Department of Construction Management & Engineering; to UK Office Systems of Liphook, Hampshire, for useful information about modern fax machines; and to those whom we have taught, in industry, and at universities and colleges, who have helped us to develop and clarify our ideas.

Joan van Emden, Reading
Jennifer Easteal, Salisbury
1996

1
Introduction

Identifying the readership or audience • Readers' knowledge and experience • Defining the objectives • Objectives and desired outcome • Writing or speaking? • Carrying out the work/Gathering the information • Checking the information

Technical communication should be a source of enjoyment as well as of information. There is a sense of achievement and satisfaction in writing or speaking effectively, and of pleasure and well-being in reading or listening to a communicator who uses words accurately, skilfully and even beautifully. The nineteenth century writer on constitutional and economic matters, Walter Bagehot, said memorably that 'Writers, like teeth, are divided into incisors and grinders'; good writing, and indeed good speaking, is incisive and also unobtrusive, allowing and encouraging the recipient to concentrate on the information. Communication of this kind demands skill and hard work, but it brings great rewards, not least in the goodwill which is built up between the writer and the reader, the speaker and the audience.

IDENTIFYING THE READERSHIP OR AUDIENCE

All good technical writing and speaking begins with the identification of the recipients and the objectives. This seems obvious and yet is too often overlooked; as a result, writing can appear self-indulgent and full of irrelevancies, and speaking can become an enclosed activity, with speakers talking to themselves and shutting out the audience. Both approaches are recipes for failure, for if the readers or audience are not involved at a very early stage, it is unlikely that they will subsequently respond well to the message—they are busy people too and they want to be sure that they are not wasting their time absorbing information which is not relevant to their needs.

READERS' KNOWLEDGE AND EXPERIENCE

The writer, then, must begin by identifying the readers and obtaining essential information about them (for the most part, what we shall discuss in this chapter applies to both writers and speakers, and so we shall use writers as the example and indicate where speaking involves a different approach). Who are the readers and users of the material? They may be easy to identify: more senior managers in the writer's company, a small number of known colleagues who need an update on the writer's work, a group of trainees who need guidance as they begin to carry out a procedure. Sometimes it is much more difficult to determine the readership: information may have to travel to a different organization or a different part of the world, and even if the writer thinks that the readership is identifiable, there is no guarantee that the report, specification or whatever, will not be passed on again and again to people of whom the writer has no knowledge.

In some ways, members of an audience at a presentation are much easier to identify. They will after all be there in front of the speaker, and are unlikely to be as varied in knowledge and experience as potential readers. They will almost certainly be there at the beginning and stay until the end; they have chosen to come because they have an interest in what is to be said. The speaker can—and must—find out how many people there will be in the audience, but inevitably the numbers will be comparatively small and they will be brought together into the same setting.

The situation of readers may be very different. They may be reading in a language which is not their own and in a cultural setting remote from that of the writer; they may have a limited interest in the document and its contents; they may be glancing through it to see if it is worth their while to read further; the information may be more relevant to their colleagues than it is to them. All this may be totally unknown to the writer, and therein lies one of the tensions of technical writing—it must be produced for a clearly defined readership and yet accessible to others. The wider the readership, the greater the problem for the writer.

There is yet another widespread difficulty. Almost all of us live with an overload of information. It comes through our letterboxes, lands on our desks, comes up on our screens, interrupts our train of thought over the telephone or the fax machine—it is increasingly difficult for readers to select what they actually need or want to read, and to have the peace and quiet in which to read it thoroughly. We therefore skim through a great deal of material, deciding what is of interest to us and also what looks accessible. We prefer a short document to a long one, and we choose what is clearly designed to meet our needs. If the writer is careless about the reader, the reader will discard the writing and may never give it a second glance—even if in fact it would have been of interest.

It is therefore very important that the writer defines the readership as far as possible, and is aware of a potential further readership which cannot be so clearly identified. There are then questions to be asked about the level of the readers' knowledge and experience of both the subject in general and the specific topic under discussion. What level of technical language is appropriate? How much detail is this major readership likely to want? How much background or explanation is necessary? The writer must decide the answers to these questions so that the information is presented at the right level, neither bewildering nor patronizing the readers. Help may be needed for some, through a glossary of terms used or appendices of supporting detail, but the target readership must be addressed in appropriate terms if it is to be encouraged to read on.

DEFINING THE OBJECTIVES

Every piece of technical writing or speaking has at least two sets of objectives—the writer's (speaker's) and the reader's (audience's). Both must be identified. The writer must ask what he or she wants to get out of writing the document: is the desired outcome to be agreement, support, more time or money, payment, the safe use of a piece of equipment; what action, from the writer's point of view, should result from the document's being read? Whatever the obvious answer or answers, there is likely also to be a 'selling' aspect to the writer's objectives. This may seem more obvious in the case of a presentation, when the speaker may be trying to impress the audience with his or her knowledge; it is also true of any document which goes outside the writer's own organization. A letter or report or specification should be telling readers that it is produced by someone who clearly has expertise in this field, who represents a well-managed and creditable organization and who is a worthy member of a respected profession. Without care, it may say exactly the opposite.

In many cases, the writer's objectives will coincide with those of the readers: the writer wants to put forward a recommended course of action to readers who want to know what should be done. This makes the writing comparatively easy. However, it may sometimes happen that the writer wants to recommend a particular course of action to readers who have already made up their minds to do something different, and who are reading the document mainly in order to destroy the writer's arguments—a much less happy state of affairs! The writer needs to know, as far as possible, what the readers' objectives are in order to accommodate them if this can be done or to undermine them if necessary.

Identifying the two sets of objectives is an essential stage. It focuses the writing, and gives criteria against which further choices can be made. The writer can decide what is relevant or irrelevant, how much detail is needed, what emphasis should be given to particular information, what can safely be left out, *only* with reference to the readers and the objectives. It is often helpful to note down the objectives at the beginning of the writing, so that they can be referred to from time to time, especially when difficult decisions have to be made. This is particularly true when there has been discussion between author and reader beforehand; a record of what has been requested and agreed is reassuring to the writer as the work progresses.

There is an additional value in having the objectives to hand. Technical writing may sometimes digress, or the proper balance of information can be lost, perhaps because some aspects particularly appeal to the writer or some information was especially difficult to obtain. The written objectives then act as a check, allowing the writer to correct the tendency to wander before further time is wasted.

OBJECTIVES AND DESIRED OUTCOME

Earlier in this chapter, we mentioned the desired outcome of the writing. This may be the same as the writer's objectives, but it may also be more precise. The writer may, for instance, wish to persuade senior managers to allow extra staff to be made available in order to complete a project; perhaps it is clear that in the long term two full-time staff need to be seconded, but the writer knows that this is unlikely to happen straight away. It remains the objective, but the desired outcome may be more limited, perhaps that one extra person is seconded immediately and the position reviewed after three months. There is no conflict between objective and desired outcome, but the latter represents a realistic short-term view of what can be achieved. The writer, or speaker, may need to identify both before starting to gather the information.

WRITING OR SPEAKING?

A decision about the appropriate form of communication may have been made long before the writer became involved; it may be self-evident (nobody would try to speak a specification or write out a telephone conversation in full); sometimes it is not immediately obvious which approach is more suitable, and sometimes the two are linked (as we shall show in Chapter 8).

In general, the spoken word is faster than the written, although modern developments such as e-mail or fax have tended to modify this view; both may be confidential (although some forms of communication, such as the videoconference or e-mail, are not very secure); the written word is always more formal than the spoken and, perhaps most importantly, provides a lasting record of what has been decided, so that it may have a legal or contractual significance. This can also be true of spoken communication, but it is less likely to be binding and people almost instinctively ask for a written record of what has been said. 'Putting it in writing' implies a recognition that the written form is permanent and decisive.

There are other, less obvious, reasons for making a particular choice. People may hesitate to put information in writing just because to do so will formalize what they have said, when perhaps they want to try it out in an informal way first. Uncertainty about how information will be received, nerves or embarrassment may all cause a retreat to writing rather than speech. Interestingly, personal dislike can also affect the means of communication, persuading the communicator to use the telephone rather than a face to face encounter, and a memo rather than the telephone.

Information may, over a period of time, change form more than once. The result of a piece of research, for instance, may be disseminated first through private conversation with fellow researchers in the same organization and then at an informal seminar, also to colleagues. It may next be given—with the permission of the company—as a formal presentation at a conference, and subsequently published either in the conference proceedings or as an article in a professional journal (perhaps both, in slightly differing formats). Eventually the researcher may publish a fuller account of the work in a book. If the subject is of general interest, it may become the subject of an article in the serious press, so that it reaches a wide range of readers with an intelligent interest but no specialist knowledge; it may, as time passes, become part of textbooks for undergraduates and even for sixth formers at school; and a brief summary might appear in encyclopaedias. By this time the researcher (no doubt very old and greatly revered) may give talks to interested members of the general public and to students internationally.

Most research, of course, does not reach this point, but it is interesting to see how such work would move between the written and spoken form. The latter would come first for reasons of speed, but very soon both forms would be used freely: presentations at meetings of experts would introduce the research to those immediately capable of understanding its implications, while the articles or textbook would reach a world-wide readership beyond the limits of the spoken word.

As the research results are disseminated in these various ways, they will change in format. In an hour's talk, it will not be possible to go into great detail about the background or even the work itself, but enough could be said about the results to interest those who probably know the background anyway. In the book, there is space for a great deal of information about what led to the research and how it was carried out. At the conference, the researcher might use a dozen slides; in the book, there could be very many diagrams, some exceedingly complex. The undergraduate textbook might have a small, select bibliography, while an article in a specialist journal might have almost as many pages of bibliography as it has of text.

Some of the major differences between the written and the spoken form are becoming clear. In speaking, we have to be selective or we should go on for far too long and the audience would cease to remember anything we said. People retain very little of what they hear. We cannot show very complex diagrams (it is difficult to absorb much information in a few minutes, especially if someone is speaking at the same time) and it is almost impossible to take in other useful information, such as a reading list. All we can do is to highlight the most important information—the results, the cost, the advantages of a particular system—whatever it might be.

The written word is very different. It can be read and re-read at leisure, a bit at a time; the complexities of a diagram can be studied at length, and a bibliography can be used intermittently over a long period. If the subject is very complex, we can study it over days or perhaps weeks, asking advice from others if necessary.

The information has therefore to be adapted, to the small amount which an audience can absorb at one hearing, to the constraints of a 5000 word article, or to the flexible limits of a book of perhaps several hundred pages. A decision has to be made each time about the most effective means of communication for the material, and it has to be adapted accordingly. The language especially has to be adjusted, either to the precision of the written word or to the greater informality of speech. It is disastrous to write like a presentation or, as we shall see later in discussing spoken communication, to speak like a report (see page 107).

Yet this does happen. Professional technical people are not always sufficiently aware of the differing amounts of detail that can be conveyed in speech and in writing, the different styles which must be adopted, or indeed the differing objectives inherent in the two forms. The technical detail can best be given in documentation; in a presentation, the speaker is convincing the audience. This may be an over-simplification, but it contains much truth.

CARRYING OUT THE WORK/GATHERING THE INFORMATION

There are many different ways in which technical information is gathered: the writer may generate some material, perhaps from test results; gather some from colleagues and other specialists; gain more by reading and listening. No method is intrinsically better or worse than any other, but all ways must be explored thoroughly. Once the readership and the objectives have been clarified, and the writer knows how the information is to be presented, the next main task is to gather whatever is relevant to the end product.

Relevance is important; later, it will become essential, but at this early stage it is better to include information if there is any chance of using it. In the next chapter we shall discuss the organization of the material and at that stage some of what has been collected will be discarded. For now, what is clearly irrelevant can be put on one side or thrown away; what is of dubious relevance can be kept.

Some material may not be available to the writer for reasons of confidentiality or the time available for the writing; this may be frustrating, but there are limits to what can be obtained before the deadline. Nevertheless, as far as possible the writer must be thorough in this stage of the work; inadequate or unsubstantiated information will simply waste the reader's time and will irrevocably damage the writer's credibility. The content of the document should be comprehensive and well-argued, and its accuracy should be checked whenever this can be done.

CHECKING THE INFORMATION

The work is being carried out by technically qualified and experienced people because of their particular expertise. For this very reason, it is easy for them to omit ideas or information just because it seems so familiar; we all tend to assume that other people know what we know, and so leave out what would in practice help or inform the reader. Informal discussion with colleagues is invaluable at this stage. They may have additional information or a different point of view; they may query what the writer has taken for granted or suggest further sources to explore. It is a sad fact that shortage of time prevents many writers from employing this easy and often pleasurable method of collecting and checking material—and much time is thereby wasted. Other sources such as the company library (or the library of a professional institution), British and International Standards, published articles and previous in-company documentation should all be available and can be used to check and double-check the accuracy of the material which has been gathered. Mistakes of fact may mislead the reader; their more lasting effect is to undermine the confidence of the reader in the writer's professional competence.

In an ideal world, all the information would be gathered before the writer moves on to the next stage; in practice, further material may come to light as the document is actually being prepared. This is a nuisance, but not too serious if there is a clearly defined structure into which additional material can be fitted. Organizing such a structure is the next task of the writer, and the first subject of the next chapter.

Key points

- Identify the target readership or audience, its knowledge and experience
- Be aware of other readers to whom your document might be passed on; it must be accessible to them

- Choose an appropriate technical language and level of detail for the main recipients, giving help to others as necessary
- Identify in detail your own objectives and those of the readers (audience), and the desired outcome of the work
- Make a note of the objectives you have identified and use them as a guide to what is relevant or irrelevant in your material
- Written communication tends to be more lasting, more formal and more binding than the spoken form. Personal feelings may influence the choice of communication method as much as rational argument
- The same information may appear in both written and spoken form; adapt it as appropriate, remembering that complex technical detail is more easily assimilated if it is in writing
- Be thorough in collecting information; at this stage it is better to have too much than too little material
- Check the accuracy of the information you have gathered; inaccuracy damages professional credibility
- Use the expertise of your colleagues as you revise the information, before and during the preparation of the text.

2
Writing Techniques

Planning the document • Organizing the information • Appropriate style for technical writing • Diagrams • Checking and reviewing • Editing group documents

PLANNING THE DOCUMENT

At the end of the previous chapter, we stressed the need to develop a structure for any technical document at a very early stage of preparation and certainly long before beginning to write.

There are two possible approaches: writing a draft document and then 'tidying it up' by imposing a structure, or planning a structure and then writing into the format that has been developed. The second method is faster, easier and more satisfactory, although it takes some practice at first for those who have previously approached their writing in a different way. Drafting a document is, of course, much simpler than it used to be in the old days when information had to be written out by hand and could not easily be moved about; part of the temptation of the wordprocessor is the ease of writing and of incorporating second thoughts. This can be a trap: we see on the screen at any time only a small proportion of what we have written, and so need a printout for revision—in other words, we are back to the problem of the inflexible printed page. By developing a structure *before* starting to write, it is possible to cut out most of the drafting and to write directly into the planned format—and thereby to save time. Revision and checking will usually be needed (see page 31), but these are less time-consuming activities than preparing a new draft of the whole.

Categories of information

The first stage of planning can take place as the information is being gathered. It needs to be sifted, at first very simply (as we said earlier, put in rather than leave out) and then more critically into three categories:

A Material which is clearly relevant to the subject of the document and which will go into the main body of the text
B Material which is borderline: it has some relevance but may not be used in the mainstream of the document—perhaps in an appendix
C Other material which is probably irrelevant, but which should not yet be discarded—the writer might have second thoughts about its relevance.

From this point, the writer will be concentrating primarily on Category A material, but will check from time to time to make sure that other information is still in the right category (the identified objectives are very useful in making such decisions). No category is likely to be complete at this stage, since information may continue to reach the writer until quite late on in the writing process.

Choice of format

The writer may have no choice in the format for the document as a whole: many organizations have outline schemes for reports and specifications available on the computer network; this saves the writer's time and ensures a certain uniformity in company documentation. There may also be an outline format for letters, faxes and memos, which are in any case conventional in structure, or earlier correspondence in a sequence of letters, which can act as a guide. The writer may well have access to previous procedures or minutes of meetings and be able to adapt them to the present purpose or simply follow the pattern of what was successfully used in the past.

All this is very helpful, as long as it does not become a strait-jacket. There should be a little flexibility, especially in a long document, so that the format can be developed to suit the writer's particular purpose. Nevertheless, the best place to start planning a format is with a previous or recommended version; this will give a useful outline but it will not show how the information should be structured in detail, and a different technique must be employed for the next stage.

ORGANIZING THE INFORMATION

There are various techniques for structuring technical documents, and perhaps the most useful comment to be made is that they are very personal to each writer; any of the recognized methods may be used or adapted, or the writer may develop a totally new technique—as long as it works for the individual, that is all that matters.

Nevertheless, many writers have been helped by some widely used methods, and in this chapter we will describe two examples. Again, there are different ways of using them, and the writer who has not met them before might like to try them out two or three times before deciding what seems to be most congenial. In each case, we will use a simple everyday example first, and then show how the technique can be applied to more complex information.

List and display technique

Simple example   Philip is going on holiday. He and a couple of friends have chosen a two-week motoring holiday, driving south (in his car) through France and spending some time on the Mediterranean coast and probably a few days walking in the Pyrenees. They are going in late September in order to avoid the main tourist season. Philip will be at work until the evening before they leave, and he knows that he will have little time for planning or packing. He decides, therefore, that he will start to gather ideas (see Fig. 2.1) well in advance of the holiday. He takes a large sheet of plain paper, turns it on its side (landscape position) and writes large well-spaced headings across the page:

Flat   Car   Packing   Documents   Travel

There are some ideas that come straight into Philip's mind. He must leave a set of keys to his flat with a neighbour and ask her to feed his cat while he is away. He therefore adds 'keys (Sue)' and 'cat' under the heading 'Flat'. He

Flat	Car	Packing	Documents		Travel	Work
keys (Sue)	service	clothes	personal:	passport	tickets	redirect phone
cat	spares	shoes		E111	money	brief (Bill)
heating	GB sticker	wash kit		insurance	maps	meeting (John)
fresh food	triangle	first aid	car:	green card	.	.
rubbish	.	cameras		RAC cover	.	.
.	.	.		MoT cert.	.	.
.	.	.		licence	.	.
.	.	.		.	.	.
.	.	.		.	.	.

Figure 2.1   List and display technique (simple example)

also realizes that his car will need its annual service before he sets off, and so writes 'service' under the 'Car' heading. Gradually the lists develop, and subheadings begin to appear, such as, under 'Documents', *personal*: passport, E111; *car*: green card, RAC cover. From time to time, other ideas occur to him: he adds the heading 'Work'—very much as an afterthought—as he realizes that one colleague will need an update on his work before he goes, and another will have to be asked to cover a meeting on his behalf.

Complex example   Philip uses the same organizational technique at work. Recently, he has assumed responsibility, as Project Manager, for a large construction enterprise, and as soon as he gets back from his holiday he will be involved in a great deal of planning. He can already see some of the problem areas. The building will be a large office block with an *in situ* reinforced concrete frame which will fill the available land. Large cranes must be used to lift panels into position, but there will be only limited space available in which to place the cranes. Space will throughout be severely restricted, and questions of storage and unloading will be a major headache. There will inevitably be some inconvenience to the general public during the construction period of 180 weeks.

All this is distracting Philip's attention as he finishes the previous project, and, in order to record his thoughts, he uses his 'holiday' technique. Again, he finds a large sheet of paper, turns it on its side, and jots down headings:

Structure   Space   Public   Equipment

He adds his thoughts to each heading: 'cranes – 6? – cost?' under 'Equipment', for example, and 'covered walkway' under 'Public'. The lists soon develop as shown in Fig. 2.2. (*Note*: This is, of course, greatly over-simplified; such a diagram would be enormously complex, involving more than one person and including many considerations such as timing, for instance. However, a few examples in each category show how the technique may be used.)

Structure	Space	Public	Equipment
in situ reinforced concrete	stores	covered walkway	cranes – 6?
.	canteen/	parking restrictions	– cost?
.	welfare	road space (for	– airspace?
.	offices	delivery vehicles)	forklift
.	unloading	.	concrete pumps
.	.	.	.
.	.	.	.

Figure 2.2   List and display technique (complex example)

Spider diagrams

Simple example   We have seen a possible method which Philip might use to organize information both personally and professionally. He might instead have used a different technique, the spider diagram (also known as mind mapping). This is a slightly different approach but produces the same result: it has some advantages in that it has less emphasis on hierarchy and more on flexibility; in the nature of a list, what is at the top is assumed to have more significance than what is at the bottom; spider diagrams have no 'top' or 'bottom' in this way. It may also be true that the patterns produced by the diagram allow our minds to move more freely over the material, seeing

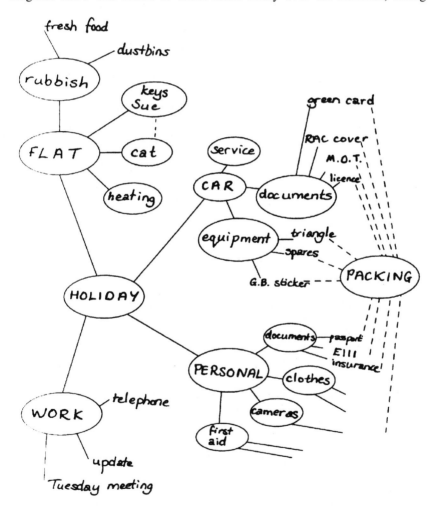

Figure 2.3   Spider diagram (simple example)

connections which would not otherwise be obvious; as we tend to remember diagrams more easily than lists of words, the information may remain in our memories for a longer time. On the other hand, some people find spider diagrams less easy to use, and practice is generally needed before the technique can be employed quickly and fluently.

Philip would begin his holiday planning by using a large sheet of paper in landscape position as before. He would write 'Holiday' in a bubble in the middle and then add his main ideas in further bubbles in a random way; secondary ideas can then be added to appropriate bubbles, or put by themselves if they do not seem to have other connections. Some ideas will have implications across the page, for instance 'passport' might be linked to 'packing' and to 'car'—so that the connection is readily visible should anyone want to look for it. These connections can be drawn in (most easily in different colours)—a feature which is one of the advantages of this particular technique.

The spider diagram will develop as shown in Fig. 2.3.

Complex example   The same technique could be used for Philip's construction project. Again, he would write the name of the building in a bubble in the middle of his paper, and the major ideas would radiate from that, each in its own bubble. There are, even in this simplified version, a good many cross-references; Philip might decide later that the diagram had become over-complicated, and take 'Equipment', for instance, as the central bubble in a new, subordinate spider diagram. This is often helpful when the writer feels that the main diagram is becoming crowded—it allows further structuring of a particular area of the information.

The spider diagram will develop as shown in Fig. 2.4, opposite. (*Note*: As before, this is a simplified, selective diagram to illustrate the technique.)

Developing the structure

Whichever technique is used, the result is organized material from which a contents list of headings can be developed, either formally, as in a report or specification, or for the writer's use only, as in an article. The main headings (the words in capitals in the bubbles) will become headings in the text, if they are needed, and can be organized into a logical structure. The subheadings (subordinate bubbles) can also be logically ordered, and in this way the whole format can be developed. Examples of such a structure in practice can be found in Chapter 4, Written Text (page 50); an article or a book could be prepared on this basis, although very little of the structure itself would be reproduced in the text.

When a constraint of length has been imposed (a two-page report or a 3000 word article, for instance), the space or number of words available for each aspect of the subject can now be allocated on the basis of its place in the overall structure and the author's own sense of what is the appropriate

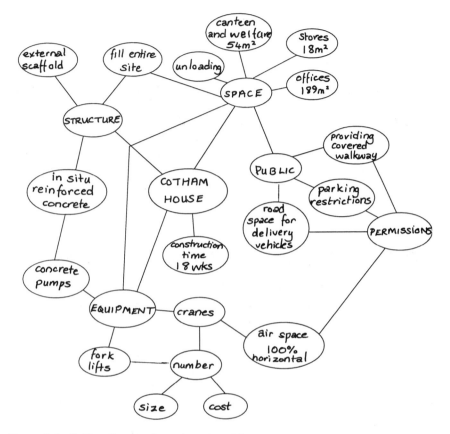

Figure 2.4 Spider diagram (complex example)

length. As the document is actually written the balance will change, but the writer will be aware of what is happening and will not need to compress or omit an important issue because the words allowed have all been used up. This is particularly important for writers of dissertations and theses, where the word limit is large but has to be observed.

By the time that the author has a detailed structure, a list of 'headings' or a 'contents list' of sections, much of the work has been completed. If possible, it is worth at this stage getting the agreement of a knowledgeable colleague, the senior manager who will be responsible for issuing the document, or, in the case of the thesis, the supervisor. Many decisions about relevance, the need for more detail, logical order, appropriate emphasis and so on have already been taken, and it is reassuring to have a second opinion before starting to write.

Getting started

Many writers hate the blank sheet of paper, the empty screen. The first paragraph of any piece of writing is likely to be the hardest; once the writing has begun to flow, it all seems to be easier and more congenial. This is where the planning has such a beneficial psychological effect: a major stage of the preparation is essentially over and the writer can therefore concentrate on the writing. Where that process actually starts is totally unimportant—any small area of the structure will do, and it is often helpful to begin not at the beginning but at the easiest place, some straightforward, factual section about which the author feels confident. It is generally a great relief to have started.

Nevertheless, the writing itself must be appropriate in technical content and style, and further decisions, therefore, have to be taken.

APPROPRIATE STYLE FOR TECHNICAL WRITING

The style in which a text is written must be suited to the readers, the information and the context. If, for instance, the reader is well known to the writer, it might in theory be possible to write in a friendly, informal style. However, if the material being conveyed is the result of highly complex technical research, it might be very difficult to put across informally, and the appropriate style would therefore have to be more distant and formal. Similarly, if the data had to be disseminated in a published article, the style would be formal, even though any particular reader might be well known to the writer.

There is as a result no clear definition of 'appropriate style' and, in considering the needs of reports, memos and specifications, for instance, we shall be commenting on the style which is conventional to each. Nevertheless, there are some useful principles to bear in mind for almost any technical material, and common problems of expression and style, and these will be discussed in this chapter. However, this is not a grammar book, although some of the problems considered are issues of grammar or punctuation; more detailed help of this nature may be found in other books, some of them listed in the Bibliography.

Active or passive?

Most technical writing is formal. The subject matter, as we have suggested above, is generally the overriding consideration; it is not usually possible to convey complex information except in an impersonal way. This means, for instance, that the writer should not be referred to as 'I' and the reader should not be referred to directly as 'you'. The writing will be as objective as possible, though not necessarily in the passive voice.

This may be seen as a break with tradition. In the past, scientific and technical writing was always in the passive, which means that the subject of the verb is at the receiving end of the action, while the person carrying out the action is mentioned only obliquely. So 'Peter carried out the experiment' becomes 'The experiment was carried out by Peter'. In much technical writing, the convention remains: many companies would insist on 'It is recommended that' and would not permit 'We recommend that'. Even this simple example shows that the sentence is likely to be longer in the passive, and that the effect is to make the style less direct; it makes less impact, indeed, it is rightly called 'passive'. The opposite, the active style, stresses the person performing the action ('We' in the example above). In some contexts, such as that of a specification, a direct, active way of writing is positively helpful in removing any ambiguity.

There is an additional danger in changing from the active voice to the passive: the meaning may be distorted. 'I cannot accept the idea' clearly shows a personal opinion; 'We cannot accept the idea' is wider, suggesting support for the writer; 'the idea is unacceptable' is definite and apparently final—nobody would accept such an idea. The personal opinion is now a statement of fact. In order to clarify the situation, the writer would need to add 'in the author's opinion', which is wordy and a little pompous. We have to be sure that in using the passive, we are not widening the frame of reference unacceptably.

Nowadays, with a general move towards less formality in both speech and writing, the active voice is becoming more acceptable, but there are often company constraints on when it is to be used. The writer should check company requirements, and under most circumstances, if in doubt, use the indirect, passive voice.

The need for simple words

'Formal' does not mean 'pompous'. Too many technical writers lose all sense of the normal English language which they would use if they were not writing a formal document and switch into an over-formal, high-flown style which they hope will be 'impressive'. This is a basic misunderstanding of the nature of technical writing, which should never draw attention to itself. If the reader thinks, 'This is easier and more interesting to read than I would have expected', the style is likely to be a good one. The reader's attention has been focused on the information, as it should be. The writing itself is unobtrusive; it does not clamour for attention, but guides the reader easily and smoothly from one piece of information to the next. Writers are mistaken if they think that it is necessarily better to 'initiate' than to 'start', and that mistakes are 'rectified' rather than 'put right'; expressions such as 'subjected to examination' (meaning 'examined') or 'we are in a position to undertake' (meaning 'we can'), are better avoided. The simple form is almost always the most appropriate. In the same way, the use of obscure classical abbreviations such as 'viz.' or 'ibid.' will not make the right impact on the reader. All that this style achieves is irritation that the reader's valuable time is being so needlessly wasted. A simple, unadorned style allows the reader to get on with the job of reading.

Using words precisely

Oddly, in technical writing it is often not the complex scientific words which cause problems, but the small, insignificant words in between. These are used carelessly because their importance is so easily overlooked. The writer of a technical article wrote that, in future, the construction industry would need less skilled workers. Challenged for the reasoning behind such an unlikely statement, he tried to justify his words—and it became clear that he had meant to write that the industry would in future need fewer skilled workers. The confusion of 'less' with 'fewer' had turned a perfectly reasonable statement into nonsense. There is, incidentally, a double error: 'less' refers to one item ('less cake'), while 'fewer' refers to multiple items ('fewer buns'); in this case the problem was compounded by the fact that 'less skilled' has its own meaning for the reader, while 'fewer' obviously relates to the unit 'skilled workers'.

A similar misunderstanding was caused by the following request: 'Let me know if the system is operational'. The system wasn't, and so they didn't, and the writer was puzzled by the lack of response; he had, of course, meant 'Let me know whether . . .'. It is important, too, to preserve 'where' for place and 'when' for time; in the sentence 'Extra resources will be allocated where the project has overrun', the reader will be unsure whether the resources are to be allocated *when* the project has overrun or at the *locations* at which the overrun has occurred, or indeed in specific *instances* of overrun.

Some words are in their nature imprecise: how 'local' is 'local'? If we know an area well, our definition is restricted to places or activities in a very small geographical region; if we are not familiar with our surroundings, we tend to see places as local to one another over a broad area. The writer may define 'local' very differently from the reader. Similarly, 'recent' or 'nearby' mean very little and should be clarified.

The effect of too many vague expressions such as 'quite', 'fairly', 'to a limited extent', or 'in due course', is to undermine credibility. The writer is seen as hedging bets, being unsure of the facts and so suspect. An appropriate emphasis can be given by relating to information which the reader knows—a price rise may be linked to the current rate of inflation; the cost of a component may be given as a percentage of the overall cost; the expected workload in a new project may be assessed in comparison to a similar completed undertaking. Many figures mean little by themselves and need to be given a context.

Meaning and implication

Words have meanings, which are generally found in dictionaries. They also have implications, which are much harder to define. We learn when to use one word rather than another with a similar meaning by listening to and using the language; this is one of the biggest difficulties in learning a foreign language. The problem is exacerbated in the case of the English language because it has so many words—often several with very closely related meanings. If one word is chosen rather than another, the dictionary meaning may remain the same, but the 'feeling', the implication, of what has been said may be changed.

This can be a problem in recording the minutes of meetings, for instance. People 'ask' or 'want to know' or 'enquire' or even 'demand to know'—and the result can sound more or less aggressive in the reading. The word 'shortage' is defined in the *Concise Oxford English Dictionary* as 'deficiency', and this is clearly correct. Nevertheless, we would be as unlikely to comment on the deficiency of pencils on the desks as on the shortage of a hundred pounds in the accounts. Similarly, we might feel the lack of promotion, but not the shortage or deficiency—yet 'lack' is also defined as 'deficiency'. The best word in any context is the one which has the correct meaning and also the same implications for both writer and reader.

Some common expressions are misused, and this can cause not so much a misunderstanding as a loss of credibility. For example, materials must comply *with* or conform *to* the specification; the machine *is composed of* or *comprises* its constituent parts. Such expressions are often confused.

Finally in this section, it is worth noting that sometimes expressions can be accidentally amusing—not an effect which is usually appropriate—and that some words are just plain ugly. A report on the management of a project included the unnerving advice: 'At this stage it is important to inject new people.' Students often do experiments and get results, and very boring it is to read about it. The words 'done' and 'got' should be banned from scientific and technical writing: experiments can be carried out and results obtained, and the reading flows and is pleasurable!

Jargon

Jargon comes in three types: professional, company and fashionable. Professional jargon is part of the stock-in-trade of every technical writer; it makes for precision and is easily recognized by fellow professionals. As long as the reader is familiar with it, there is no problem (but see the use of the glossary in Chapter 4 under report writing, page 55).

Company jargon develops in almost all large organizations, and its use internally causes no difficulty. However, as time passes, staff become so familiar with company terminology that they forget that the rest of the world speaks a different language, and that can be a major problem. All technical writers must be able to identify company jargon and to use it with great care, and only to appropriate readers.

Fashionable jargon is much more insidious. Just because it is widespread, people are unaware that they are using it; if the habit takes hold, the effect can be distracting and irritating. To work in the engineering function means to be an engineer; to be in an in-company situation is to be in-company; 'at the end of the day' means not in the evening but eventually or finally. Such expressions have their place but, if they are used frequently, they attract attention to themselves and away from the message.

Hyphenated words

We use much less punctuation than in the past (see page 24). The widespread use of wordprocessors has resulted in very few word-splits, and it is assumed that readers will know instinctively when words should be connected. However, much technical material is produced for readers in other parts of the world who do not necessarily group the words correctly at first reading. Indeed, most readers would have to re-read the following sentence in order to understand it:

The insulation is made of glass reinforced foil faced mineral wool.

Two hyphens greatly improve the ease of reading:

The insulation is made of glass-reinforced foil-faced mineral wool.

Many such expressions will benefit from the hyphen, for example, colour-coded, hot-dip, fire-resistant, sound-absorbent; few nowadays would be written in this way. The disappearing hyphen should be reinstated in the interests of clarity and reader comfort.

Sentence structure

Two major problems confront many technical readers: sentences are commonly too long, and writers, being unsure about punctuation, leave it out. Both tendencies show themselves in reports, specifications, technical letters—in fact, in most of the documents considered in this book. There is no space here to deal with all the related problems in detail, but this section will look at some of the most common effects of the over-long sentence and missing (or peculiar) punctuation.

When readers are faced with a passage of densely technical material, they look anxiously for a point at which to pause, in order to assimilate what they have read so far, and to draw breath for the next onslaught. Such a pause is signalled by a full stop. If they have to read 50, 80 or even 100 words (such sentences are to be found in many technical documents) without such a respite, panic sets in; they will then either give up completely and skip ahead to find something more palatable, or they will put in their own punctuation, taking the risk that in doing so, they will change the meaning of the passage. From the writer's point of view, neither course of action is desirable.

Sentences become too long for a variety of reasons: the writer is excited by the subject, and information pours out in a chaotic way; the writer is thinking his or her way through the subject, and includes the whole thought process without pausing to consider how much the reader needs to know; occasionally, the writer is dictating the material and loses track of the sentence length; more frequently, irrelevant ideas creep in almost as afterthoughts, and are included more or less by accident. Perhaps most commonly of all, the writer is seated at the wordprocessor and allows words to appear on the screen as if by magic, with no sense at all that anyone might need to read them.

The following real-life sentence shows the effect of such processes:

> After careful consideration of price, delivery and the company's expertise in the manufacture of consoles, it was the unanimous decision of the group to recommend Joe Bloggs as the subcontractor best suited to carry out construction of the consoles to our design, subject to their justification of their tooling charges, since the detail of the information being given showed these to be considered excessive, to this end Jim Smith was requested to set up a meeting with them so that this matter could be discussed.

Much of this 85 word sentence is redundant, and what is left is verbose and unstructured. It can be analysed as follows:

> After careful consideration   It should surely be taken for granted that the group would take an important decision such as choosing a subcontractor only after careful consideration. Why say so, unless there is any doubt on the matter?

> of price, delivery and the company's expertise   The choice of subcontractor is likely to be based on factors such as price, delivery dates and known capability. Again, why say so?

> it was the unanimous decision of the group   If the decision was taken, why should it matter whether it was unanimous or not? There is a slight suggestion of something hidden—had the group been seriously divided over previous decisions?

> best suited to carry out construction of the consoles to our design   Joe Bloggs is the chosen subcontractor. Would he have been chosen if there were someone better? He will naturally be expected to use the client's design.

> since the detail of the information being given   This is almost meaningless. What information, given to whom? Presumably that given to the group to help them decide.

showed these to be considered excessive Who considered them to be excessive—the group, or some other unknown influence? Excessive in comparison with what?

to this end This presumably refers to 'their justification', but it is a long way away in the sentence, and is unclear and unnecessary.

was requested to set up Perhaps he was asked to set up a meeting, or perhaps he was simply asked to meet Joe Bloggs, which seems more likely.

so that this matter could be discussed This is the purpose of the meeting, which has already been planned; the words duplicate 'to this end', and could be left out.

When all this unnecessary material is omitted, we are left with three small pieces of information:

the group recommends Joe Bloggs as the subcontractor to manufacture the consoles

Joe Bloggs's tooling charges appear to be too high

Jim Smith has been asked to discuss this with Joe Bloggs

This could be left as three sentences, but the order is unhelpful. The casual reader might see the recommendation, and read no further (this, of course, was even more likely in the original version), thus being left with the wrong message. It might be better to say:

Joe Bloggs is recommended as the subcontractor to manufacture the consoles, but his tooling charges seem high. Jim Smith will discuss these charges with him, and provided they reach an agreement, Joe Bloggs's tender will be accepted.

The 85 word sentence is reduced to two sentences totalling 37 words, and the order in which the information is given is clearer and less likely to mislead.

The average length of a sentence in densely technical writing should be between 17 and 20 words. This is quite short, but as it is an average, it allows for some variation—as good style requires. Some sentences can be as short as five or six words, while others can be correspondingly longer, but the sensible maximum is about 30 to 35 words. This is short enough to allow the reader to look ahead and to see the place to pause for the assimilation of ideas; it is also an encouragement to read on. Technical information should be absorbed in small chunks, so that it is not only read but understood and remembered.

There are other advantages in the control of sentence length. The very short sentence attracts attention to itself—it jumps out from the page. The skilful writer will therefore use such a sentence to highlight a key message: 'The new process will save money', 'Such a change will benefit the company', 'This is the maximum temperature permitted'. The short sentence will probably be followed by rather longer sentences giving more details.

Within any text, there are likely to be opportunities for sentence variation: the short sentence might be used at the end of a paragraph to sum up the main point; a list of facts (see page 27) may contain several short messages; a passage which calls for an assessment of the implications of the data might be written in rather longer sentences, which slow the reader down and encourage reflection. Readers are guided through the document—probably unconsciously—in part by the variety in sentence length.

Sentences should not be too long; they should also be well-constructed. A lengthy digression between subject and verb is difficult to follow: by the time readers get to the main point, they have forgotten how the sentence started and have to read it again—and inevitably feel irritated with the writer. In the following sentence taken from a company report, the examples distract from the key message:

Outfalls, for example those from power stations, sewage works, chemical and other works, also river estuaries, can be detected by the disturbance they cause to the surface of the sea.

Between the subject 'outfalls' and the verb 'can be detected', there are 15 words, forming five examples (power stations, sewage works, chemical works, the unidentified other works, and river estuaries). The reading is not helped by the fact that 'works' occurs twice, and that the comma followed by 'also' is distracting. Are the river estuaries simply another similar example, in which case they should be joined by 'and', or are they a different group in some way? We may reasonably decide that they are, being natural phenomena while the other examples are man-made. By the time the reader has reached 'can be detected', there have been too many distractions, and the sentence will probably have to be read again.

If so many examples are really needed, it would be better to put them into a separate sentence:

Outfalls can be detected by a disturbance to the surface of the sea. Power stations, chemical or sewage works, or natural phenomena such as river estuaries, may be identified in this way.

A further real-life example shows the need to plan sentences and to be aware of the unfortunate impression which can be created by unstructured material:

The risk from Legionnaires Disease and the advantages of taking plant out of areas which could be more favourably utilized has prompted this investigation into air cooled versus water cooled refrigeration plant.

The writer is considering a particular project, a comparison of air cooled and water cooled refrigeration plant, and this appears to be the main point of the sentence. Why was the project started? There seem to have been two reasons, the major one, of universal concern, being the risk of Legionnaires Disease. In the case of this particular investigation, there is a secondary motive, of local concern only—the need to save space. The writer shows no awareness of the difference between the two motives, and gives them equal emphasis. This tactlessness, combined with the back to front structure of the sentence (the main point comes in the second half), sounds clumsy and oddly callous. It might have been better to stress the organization's main concern—the well-being of its workforce—by leaving out the second motive altogether:

> This comparison of air cooled and water cooled refrigeration plant was prompted by concern about the risk from Legionnaires Disease.

Back to front sentences are very common. The writer analyses the message and writes it in the same order as the analysis; the reader wants to know what the main point is, in order to understand the significance of the detail. The problem is illustrated in the following sentence:

> In order to ensure that the process can continue in spite of the need for a substantial overhaul of the machinery, new procedures will shortly be issued.

Readers of this message realize that *something* will happen in order to ensure that the process can continue, but they have no idea what. Is this a message about increased manpower or new shifts or additional machines? They will not find out until the end of the sentence—new procedures are to be issued. If the order of information is reversed:

> New procedures will shortly be issued to ensure that the process can continue in spite of the need for a substantial overhaul of the machinery.

the ideas can be assimilated in a logical order, and the sentence will be understood at a first reading.

Punctuation

We use much less punctuation than in the past, as we noted earlier in talking about the need for hyphens. This is in some ways a good thing: it avoids the cluttering of text with a great deal of 'heavy' punctuation, and of course it makes wordprocessing very much quicker. If carried to excess, it may make the reading very much slower.

The basic punctuation of capital letter at the start of a sentence and full stop at the end is generally observed, although technical writers are not always sufficiently mindful of the rule that a sentence must contain a main verb and must make sense by itself.

> Bearing in mind that the Building Regulations stipulate an absolute minimum width of 800mm for external doors.

is not a sentence, as it fails to meet either requirement. A main verb, for instance 'fulfil' needs to be added, and the reader must be told what is the effect of bearing the Regulation in mind. As it stands, this message is incomplete. It might sensibly read:

> Bearing in mind that the Building Regulations stipulate an absolute minimum width of 800mm for external doors, our plans fulfil the requirement.

The comma after 'doors' is not essential but it is helpful to the reader, in showing that the first part of the sentence has been completed and that the main point is about to follow. An additional comma would be used if the sentence were to continue with a further point:

> Bearing in mind that the Building Regulations stipulate an absolute minimum width of 800mm for external doors, our plans fulfil the requirement, although an ideal width would probably be closer to 1000mm.

This sentence is now carrying just about as much information as it can, and a new sentence might be preferable if yet more information is to follow:

> Bearing in mind that the Building Regulations stipulate an absolute minimum width of 800mm for external doors, our plans fulfil the requirement. However, an ideal width would probably be closer to 1000mm, and our clients might like to reconsider the plans accordingly.

The comma after 'however' is another example of helpful punctuation: the word can be used, as here, simply to form a link with what has gone before (this is a fact ... however, there is another fact to hold in tension with it); it can also mean 'in whatever way', as in 'However we look at the problem, we can see no solution.' In such a case, the sense demands that there is no comma break between 'however' and what follows. If we read the two examples aloud, we can hear the difference.

'Heavy' punctuation such as brackets should not be over-used; brackets interrupt the flow of the reading and should be kept for material which is not an integral part of the sentence, such as a reference to a diagram or an appendix. Semicolons should also be used sparingly, to join sentences which have related subjects, such as :

> Digitized speech provides good quality sound. Its limitation is that the computer can speak only the words that are recorded.

which may, if the writer wishes to emphasize the connection between the two points, be written as:

Digitized speech provides good quality sound; its limitation is that the computer can speak only the words that are recorded.

At the other end of the scale is 'light' punctuation, which may be informal and casually used. In writing a note to a friend, we may use the dash—or pairs of dashes—frequently, and the dash obviously has its place. However, the danger of using it as punctuation in technical material outweighs any advantage; the following example was found in a specification:

Cold water test – 12°C for 90 seconds.

Only a similarly structured message about a hot water test alerted the reader to the fact that minus 12°C was not intended; the note should have read:

Cold water test at 12°C for 90 seconds.

The presence or absence of a comma may change the meaning of a sentence, and the technical writer must always be aware of the possibility of confusion, as in the following example:

In particular drain pipes . . .

which should have read:

In particular, drain pipes . . .

The reader can 'correct' the reading later in the sentence, but the flow has been interrupted and the sentence might have to be read again. More seriously, missing punctuation can cause ambiguity, as in the following:

Iron must be used where extra strength is required or for exposed pipes inside buildings spun grey iron or ductile iron should be used.

This could mean:

Iron pipes must be used where extra strength is required or for exposed pipes inside buildings. Spun grey iron or ductile iron should be used.

That is, spun grey iron or ductile iron should be used in all cases. It could also mean:

Iron pipes must be used where extra strength is required, or, for exposed pipes inside buildings, spun grey iron or ductile iron should be used.

This would suggest that spun grey iron or ductile iron should be used for exposed pipes inside buildings but not necessarily where extra strength is required. Readers must decide for themselves—a dangerous situation to be in.

Accurate punctuation is essential. We have commented upon some of the most common problems, and in the Bibliography have suggested books which will give further advice on this difficult subject.

Paragraphs and lists

Paragraphs follow much the same rule as sentences: if they are too long, the reader will be discouraged and possibly confused. In theory, a paragraph should have unity of theme; in practice, it is often better to break a paragraph into two and use a word or phrase to show the link than to allow a single paragraph to expand to fill half a page of text.

There is a problem with over-long paragraphs which is especially acute in technical writing. If a paragraph contains numbers on two or three lines, it is easy for the eye of the reader to slip from one line to the next, and for the information to be misread. In this case particularly, three paragraphs to a page is a sensible minimum, although naturally the sense will sometimes make a longer paragraph unavoidable.

In many cases, the best alternative is a list. Much information can be given in list format, with advantages to both writer and reader. The writer establishes how many factors have to be given, and mentally checks them off as they are written, while the reader has an overall view of the number of factors involved and can literally mark them off on the page as they are dealt with. For this reason, instructions and procedures in particular should be given in list form, one stage at a time (see page 75).

There is another advantage of the list: it creates space on the page. A list down the page looks attractive—much more so than a dense paragraph of text, and the reader is therefore encouraged to read on.

Lists may be given in either of two forms. In some cases, the actual order in which the items are given is unimportant (as in the case of a list of equipment, for instance); in this form, the individual points may be identified by an asterisk or a 'bullet' at the start. The common practice of using a small dash is not recommended, as the dash can so easily be read as a minus sign or a hyphen.

In other lists, the order is important—the steps must be taken in a particular order for the process to be completed satisfactorily. As a result, the stages should be numbered (or lettered) to reinforce the order in the reader's mind. Using roman numerals is not a good idea, as they are easily mistyped (and are harder to check than arabic numerals), and after 10 are harder to read.

The following example shows the information which should be given in an impact test report. If it were written in a paragraph, it would be difficult to follow, and the reader might well leave out one of the points. The order has some importance, as it suggests a conventional format for such a report, and so the list is numbered.

The following information is given in an impact test report:

1 type and description of material to be tested
2 quantity of the batch from which the test pieces were taken
3 number of test pieces
4 test temperature
5 mass of the striker
6 fall height
7 total number of blows
8 true impact rate

This list is seen as consisting of one sentence, starting with 'The' and finishing with 'rate', which is why there are no capital letters at the beginning of the items. If the individual points were too long to fit into this format, it would be sensible ·to make the introduction into a separate sentence ('An impact test report contains the information listed below.') and then to start each item with its own capital letter. A final full stop might be added after 'impact rate' to emphasize to the reader that this is the end of the list.

It is essential to be consistent in a list. If the format suddenly changes (for instance, if one item in the list above were written as a complete sentence while the others remained short phrases, or if a comma suddenly appeared halfway through) the reader assumes that this is a new list, and tries to find justification for this. If there seems to be no reason for the change, the reader decides that it is one list after all, but by this time the flow of the reading has been broken and the reader has been needlessly distracted.

Linking words and phrases

As we read a text, we are to a certain extent thinking ahead. We assimilate the information which is the focus of our concentration at a particular moment, but we are also looking for a context, to see how *this* piece of information fits into the general scheme of things. Small and apparently unimportant words and phrases help the reader to see the way in which the words form a pattern and therefore to read on with confidence, subconsciously prepared for the next stage in the argument.

Such links guide readers through the detail, allowing them to create a picture from otherwise discrete pieces of information. The connection may be simple (first, secondly, thirdly; also), or it may be more complex. 'On the other hand', for example, tells the reader that the message which follows has to be held in tension with what has gone before; it is not a contradiction ('in contrast') so much as a different point of view, and the reader may be asked to choose between them. 'Therefore' implies a clear logical connection, and should not be used otherwise. If these expressions are left out, perhaps for reasons of brevity, the reader has to make the connections and may frequently have to go back a few lines in order to understand how the details are related to one another.

Good style, as we have said, should be unobtrusive. If readers stop to think about the difficulty of reading, or about the number of times they have to re-read, they will be distracted from the essential message; if the material is highly technical, they may be put off altogether, not by the complexity of the subject but by the problem of concentrating on the details when the reading itself is painful. Good writing is not only free from grammatical or other mistakes; it flows smoothly and logically, and encourages its readers to read on with understanding.

DIAGRAMS

The old adage that a picture is worth a thousand words is as true in technical writing as it is elsewhere. Information which appears complex and difficult to interpret may be made clear at once in a diagram; part of the planning of a document is to ask at what points the reader is likely to be helped by seeing an illustration which clarifies the text. As this is the primary purpose of diagrams, it is obvious that the information given in the text must be supported by that in the diagram—the two must never conflict or appear to contradict one another.

The reader needs to be told when to look at the diagram to obtain maximum help from it. This seems obvious, but if the diagram is not available in the right place, or is not clearly linked to the text, it may be ignored. Its position in the document is therefore important. Diagrams should in principle be placed where the reader needs to see them, that is, on the appropriate page of the text if possible, or on the page facing the text, and at the end of the document only if the diagram itself is an 'optional extra' which most readers will not want to see. We have probably all experienced the frustration of working with a text which is clarified by a diagram 20 pages further on, so that we either turn backwards and forwards through the document or we give up and ignore the diagram.

Sometimes, however, the best solution—putting the diagram on the same page as the relevant text—is impossible. If the diagram is too large and has to be folded into A4 shape, it will need to go at the end; this is also necessary in the case of a very long illustration, for instance a table which occupies several pages and which therefore seriously interrupts the flow of the reading. For this same reason, a diagram should never appear in the middle of a sentence, so that the reader 'jumps' the diagram to finish reading—and perhaps fails to go back to consider the diagram itself. The writer must choose the best position for the diagram, and as with all aspects of technical writing, the convenience of the reader is the major consideration.

A diagram must be referred to in the text, so that the reader's attention is drawn to it at the appropriate moment. The most common form of reference is to use the word *Figure* followed by the number of the major section in which the diagram appears, followed by a sequential number. Figure 3.6 is therefore the sixth diagram in the third section of the document. This information is often put in brackets to separate it from the main text. Tables are sometimes numbered separately from other diagrams, but with this one exception, figures of all kinds should be given in one sequence. (Some documents, of course, such as articles or books, may not have numbered sections, but in such cases the editor or publisher will explain the house style.)

Each diagram has both a number and a short title. This information should be consistently placed below the figure itself, as should keys, explanations, and the source if appropriate. None of this supporting material should appear within the framework of the diagram itself, as space in a diagram is itself giving information. The visual impact, for instance of a line near the bottom of a graph, is important, and the reader may be misled if the resulting space is filled by comment.

Within the diagram, all labels must be placed horizontally for ease of reading; abbreviations, symbols and so on should conform to International Standards as they do in the text. Consistency in this is essential, so that the connection between diagram and text is clear to the reader.

A technical writer nowadays can make an informed choice of format by trying out on the computer different ways of presenting the information. It is worth remembering, however, that all computer packages do not produce good figures, and some decisions are in any case still those of the author. An appropriate format has to be selected, for instance a table will give detailed, accurate data, while other charts will emphasize the analysis—a graph may be used to show a trend, or a barchart to show changes. Both photographs and three-dimensional line drawings can be useful in illustrating a piece of machinery, and a procedure may best be shown by a flowchart. The convenience of the reader is the principal consideration, and the writer will already have determined as closely as possible not only what the reader needs to know but how much detail the reader wants—there is no point in giving the results to six decimal places if only an indication of the trend is needed.

Diagrams, to be effective, have to be well presented. Small, cramped diagrams spoil the appearance of the text and are difficult to use. Space is essential, not only, as we have said, within the diagram itself but also on the page where it appears, so that it stands out clearly and none of its accompanying information can be confused with the written text. The quality of any photocopying must be good, and it is worth remembering that tables and charts do not fax easily: detail may be lost.

Most readers are attracted by diagrams. In choosing whether to read an article or a book, many people flick through the pages quickly and pause to look at the diagrams. If they appear to be clear and helpful, there is an assumption that the writing will be the same. The reader may have no choice about reading other documents, such as a report or a specification for instance, but the frame of mind in which he or she approaches the task may well be influenced by a preliminary look at the diagrams.

CHECKING AND REVIEWING

In this section, we shall consider three stages in issuing technical documents: the checking stage, carried out primarily by the writer; the review of the document, generally by a more senior colleague; and editing, in the limited sense of bringing together either sections written by different people or a series of documents with a range of writers. All these functions are, of course, not appropriate for all technical material, and perhaps the first decision is how far they should be carried out in any particular case.

Every document should be checked. In the case of a memo to colleagues, such checking is unlikely to be more than putting the page through the computer spell-check and quickly reading it through. This may be enough for many internal documents, depending on the technical content; if there is a problem, it is likely that the recipient of the document will walk across the office or pick up the telephone to sort the matter out.

However, there are three categories of technical material which need much more systematic checking:

1 any document which has implications for health and safety, or the well-being of employees or visitors to the company
2 any document, such as a procedure, which affects the smooth running of the company, of its machinery or its systems
3 all documents which go outside the company to clients or customers (present or prospective) and which are in a sense 'selling' the good name of the company.

Very many documents are covered by at least one of these categories, from the business letter, which will be in the third category, to the technical specification, which may be in all three. Yet the checking process is too often overlooked or treated casually; managers put forward time—and therefore money—as an excuse for negligence in this stage of document preparation. They underestimate the time wasted by errors, in telephone calls or meetings which are called because of uncertainty or misunderstanding. Even more seriously, they fail to recognize the loss of credibility, the point at which the reader thinks 'I am glad that my life doesn't depend on the accuracy of this document, because I don't trust it any more.' This is especially true of numbers: if two letters in a word are transposed, the reader will probably recognize the word; if two figures are transposed, the reader is unlikely to see the problem or know how to put it right. It may be unfair to judge information by the way it is presented, but readers do in practice react badly to material which looks slipshod, however accurate the information may be.

The three categories given above will probably be put through a more or less rigorous checking process, which might include the following stages, given as areas of responsibility.

The writer

1   The author reads the text on the computer and puts it through the spell-check. This stage is often neglected because of the irritation caused by the computer dictionary failing to recognize many accurate technical words. If they are likely to be used often, they should be added—*accurately*—to the dictionary.
2   A printout should be made but left untouched for as long as possible— 48 hours is sensible, if it can be managed. At the end of this time, the writer reads slowly through the text, pausing after each 15 minutes or so. It is helpful to cover each page with a blank sheet of paper and to reveal and check one line at a time; concentration is increased, and the writer is less likely to be distracted by what is further down the page.
3   At this stage, the writer will make any amendments or carry out any obvious rewriting.

Two copies of the corrected document should now be made and given to colleagues. These copies should include diagrams, references and other additions to the printed text.

Colleague 1
This colleague is familiar with the topic but not as knowledgeable about the detail as the writer. His or her task is to look for technical errors, mistakes of fact, ambiguities, errors of logic, too little or too much explanation. Diagrams should be checked as thoroughly as words.

Colleague 2
The second colleague will know much less about the subject, and has the primary responsibility of checking the text by looking for typing errors, mistakes of grammar or punctuation or problems of layout; at the same time, it is helpful if this colleague also looks out for ambiguity or lack of explanation.

The writer

The text, with amendments, will now go back to the writer, who will correct (if necessary with further consultation) and rewrite as appropriate. The text must of course be put through the spell-check again and a new printout checked. This is an essential stage, as corrections themselves may create new mistakes. It is always sensible to check the context of a correction.

Senior manager

A more senior manager is likely to be asked to review and countersign the document before it can be issued. This provides an extra level of checking, but the primary concerns are likely to be company policy and the relationship with the client: any breach of regulations should be picked up at this stage. The senior manager will also look at the tone of the whole, in the context of past experience of the recipient, to ensure that the document reinforces goodwill between the parties.

The writer

The writer may now have a final opportunity to go through the text in the light of the senior manager's comments and to make last minute changes, after discussion if it is needed. This is also the last chance to check the text, especially where corrections have been made. The date on the document should reflect this final revision, as the writer is not responsible for changes in the law, or prices, or technology, which take place after the document has been signed and issued.

This pattern of checking—or similar procedures such as checking by members of a group—is time-consuming and expensive. Nevertheless, it will result in a trustworthy document which is a good ambassador for the individual and the company which produced it.

As we have stressed, discretion is needed as to whether such rigorous checking is worth while; every organization should have a checking policy for different levels of document, and every individual writer should attempt to produce an accurate text. It is, however, a sad fact that in spite of the most careful checking, few documents of any length are perfect: there may even be a printing error on this very page!

Increasingly, there are technological aids to the business of checking. Grammar checks have some use in warning the writer of problems such as the over-long sentence, but if they are used indiscriminately, they will produce characterless, bland writing which seems to belong to a machine rather than to a human being. Words may be checked in context, and the computer's thesaurus may suggest useful equivalents to over-used words. No doubt all these aids will improve in time, but they should be used guardedly at present, not least since they tend to reflect American rather than British style.

This is also true of the most useful aid, the spell-check, which is invaluable for picking up typing errors. It will, as its name suggests, also pick up some spelling mistakes, but it is not always as wise as a human being. It is all too easy to create the wrong word, or to use the wrong form, for example 'principle' for 'principal'. Such mistakes may be irritating although sometimes amusing; if they occur too often, the sentence may look incomprehensible. The extreme to which this may lead can be illustrated by the following sentence, which was totally acceptable to the authors' spell-check. Readers will realize at once that almost every word is wrong:

Ewe mite sea a ruff draught inn too daze thyme butt eye no yew wood weight four a weak or sew.

EDITING GROUP DOCUMENTS

Many in-company documents are produced by several people, each supplying a section or sections according to the individual's expertise. As a result, the text may read as if it has been produced by a committee, with everybody making his or her own mark. A lack of consistency undermines confidence, and the reader may also feel that nobody has been responsible for overall checking. The inconsistencies may themselves be trivial—organise mixed with organize—but the effect on the reader is similar to that produced by typing errors: can we trust this document?

At an early stage of preparation, one member of the group should be given editorial responsibility. Some decisions can be made at once: is the spelling to be American or English? Are units to be metric, imperial or both? Are diagrams to be integrated into the text or put at the end? What British Standards (Eurocodes etc.) are to be followed? What procedures apply? How is the text to be produced—on a desk-top publishing system, an individual PC, by a commercial printer? Are all the systems in use compatible?

Perhaps surprisingly, these preliminary questions are not always asked, or are asked too late in the production of the text. The answers may not be passed on to everyone who needs them. (We have heard of an instance in which a decision about the language in which a text was to be published was made and circulated; subsequently there was a change of plan, which was passed on to most contributors but not all, so that one unfortunate wrote his part of the text in the wrong language.) As decisions are made or changed, it is essential that the information is circulated to all the group; it is much easier to write to a particular standard than to have to rewrite later.

Some companies have style manuals, in which policy issues are set out and a checklist provided to help editors who have to make their own decisions. This is helpful, especially as such a guide can itself be updated and reissued from time to time. Nevertheless, it cannot cover all eventualities, and the editor will have to make other decisions and notify colleagues—and at the checking stage, make sure that the whole document is consistent.

There are many such minor decisions. Recently, we were asked to check a short report based on the findings of some half a dozen academics in the field of construction; below are some of the questions that had to be answered.

1 How are diagrams labelled? Are they Figures, figures or figs?
2 Where is information such as the key or the source positioned?
3 How are hyphens used, for instance in co-ordinate (coordinate), up-to-date (up to date), start-up (must be consistent with sign-off), back-up (back up, backup), long-term (long term), on-site (on site, must be consistent with off site), sub-contractor (subcontractor)?
4 How are 'new' words such as word processor (wordprocessor), data base (database), on-line (online) presented?
5 If Latin expressions such as in situ (insitu) or inter alia are used, are they to be in italics?
6 Is s or z used in words such as organise (organize), organisation (organization)?
7 How are lists presented? Are they numbered or lettered?
8 What is the format of headings and subheadings?
9 Are all abbreviations used consistently? How are they defined in the text?
10 Are abbreviations such as e.g. (eg), i.e. (ie) punctuated?

Some of these decisions affected others: if, for instance, the more modern form 'coordinate' was chosen, then 'subcontractor' and 'eg' seemed more appropriate. People sometimes have very strong opinions about these questions, and the editor may need outside support, for instance from the usage in professional journals or recently-published textbooks.

Fashion or upbringing will dictate many of the editor's decisions, and in themselves most of them are unimportant, but the impact on the reader is significant: the document should look as if it has been produced by one organization working with care and attention to detail, and so earning the goodwill of the recipient.

Key points

- As information becomes available, allocate it to the main text, the appendices or a 'reserve' file; this organization may change as the overall structure becomes clearer
- Use previous formats as a guide; follow standard company formats if they are available
- Choose an appropriate technique for organizing the information in detail; be ready to adapt it to suit your own needs

- If you have a word limit, allocate an appropriate number of words to each section before you start to write
- Get agreement for the structure of your document as early as possible
- Start writing where it is easiest for you; if you have a good outline to work to, the order of writing is immaterial
- The style of writing should be suited to the readers, the information and the context
- Follow company requirements in the choice between active or passive; if it doesn't matter, you may well find that the active voice makes a stronger impact
- Use a simple, unadorned style; the writing itself should be unobtrusive
- Be precise in your choice of words, consider their implications and use them with care
- Avoid unnecessary or irritating jargon
- Use hyphens and other punctuation to clarify the meaning
- Sentences must be kept under control; vary the length but do not let any become too long
- Give the information in a logical order, putting the main point first (most of the time)
- Lists are more helpful to the reader than long paragraphs
- Show the relationship of one idea to another by the use of linking words and phrases
- Diagrams should support the written text and not conflict with it
- Consider the convenience of the reader in choosing where to place the diagrams
- Every diagram should be clearly identified and labelled
- Diagrams should be appropriate in format and well-presented
- Every document should be checked as thoroughly as its importance warrants. Involve your colleagues in this process if you can
- If a document is produced by a group, someone must take editorial responsibility, circulating information about decisions to everyone who will be affected by them. Consistency is essential
- A document which is well structured, carefully written and thoroughly checked will create goodwill on the part of its readers.

3

Technical Correspondence

Modern formats • The facsimile (fax) • The letter • Electronic mail (e-mail) • The memo

MODERN FORMATS

There have probably in the last few years been more changes in the field of technical correspondence than in any other area of communication. Such developments are by no means over, and we must expect people's attitudes to shift again and new conventions to appear. The influence of technological change will quickly make some forms of communication obsolete (as the telex has largely been superseded by the fax) and the possibilities of others will dramatically increase (for example, videoconferencing allows meetings of colleagues who are physically at a distance).

Nevertheless, human beings are hesitant about change and sometimes unwilling to put aside the habits of a lifetime. The advent of the 'paperless office' was hailed as a great space and time saver when computers first became commonplace; in fact, the generation of paper has become a hallmark of the computer society and filing cabinets are fuller than ever.

However, the uses to which particular forms of correspondence are put have changed, and in this chapter we will look at some of the traditional media such as the letter and the memo, and at some of the newer, such as the fax and e-mail. The relationship between them is an interesting one.

The most common and most formal type of business correspondence has traditionally been the letter. In the past, it had rigid conventions which have been undermined by technology, for instance, the punctuation of addresses has almost disappeared—it is simply easier and faster to wordprocess without these additional keystrokes. A punctuated address now has an oddly old-fashioned look to it, and may suggest an old-fashioned approach to work; a modern company wants to give the impression that it moves with the times.

Again, in the past, the most common informal type of written correspondence has been the memo. It has been used in companies as a comparatively quick and simple way of sending brief messages to colleagues, usually within the same building. Often no record of the memo was kept, but as it rarely contained information which was not ephemeral, this was unimportant. However, in recent years a tendency to keep duplicates of memos has developed; this has made them slightly more formal, and led to a huge increase in the paper around the office.

In the last 10 years or so, these forms of correspondence have been challenged by, respectively, fax and e-mail. Speed of communication is of paramount importance, followed perhaps by confidentiality—a letter is reasonably secure but very slow; the fax is not particularly confidential, but very fast, provided that the receiver is on hand to receive it. E-mail is fast (with the same proviso) but not secure.

All these forms of correspondence have also been undermined by the much greater use of the telephone, which has in a way become the front line form of communication. It is extremely fast and on the whole reasonably secure (perhaps with the exception of some mobile phones); it operates world-wide and is easy to use; almost every household—and certainly every business—in the developed world has a telephone.

What, then, are the main uses of the written forms of correspondence today? We will look in turn at each form, assessing its place in modern technical communication and illustrating its conventions in so far as they exist—in some cases, there has not been time for such conventions to develop, and as a result the form may be seriously misused.

THE FACSIMILE (FAX)

The use of fax machines has increased enormously since they were first introduced. They rapidly challenged telex and, being small, comparatively cheap, and easy to use, they won the race for general popularity. Among consultants, freelance writers, small businesses and so on, the fax is particularly invaluable; it has become cheaper, smaller and better quality, especially in the use of plain A4 paper, which is firmer to handle, produces a clearer image and is easier to file than the thermal paper with which faxes started and which is still widely used.

Advantages and disadvantages

The great advantage of the fax, as opposed to the telephone with which it is usually combined, is that it conveys documents; moreover, there is a written record of what has been said or the diagrams that have been transmitted. It is very rapid in transmission, and the sender is reassured that the message has indeed got through to the number dialled. There is, of course, no guarantee that it has reached its intended recipient, and sometimes the slowest part of the fax's journey is not from one part of the world to another, but down the corridor from one office to another!

The attendant constraint is that there is little confidentiality, as the message can be picked up and read before it has reached its destination. This problem will be overcome to a certain extent as more and more individuals have their own faxes; for the present, one way of achieving confidentiality is to telephone the recipient with the warning that the fax is on its way. Unfortunately, this undermines one crucial advantage of the fax, which is that it can be transmitted overnight at the cheapest telephone rate, to be found first thing in the morning.

Such a lack of security cannot be overcome by messages on the fax itself: one of us received a fax intended for someone else, and, tearing off the long scroll of paper, started to examine it, working slowly backwards through the message until the first sheet was reached—at which point the intended recipient's name appeared, along with the message 'CONFIDENTIAL'. Nevertheless, information can be sent from one company to another in the reasonable security that it will not be seen outside the organizations concerned. Some modern fax machines incorporate a confidential reception facility, the message being held in the memory until the authorized recipient enters a password to initiate the printing. This is a welcome development.

Faxes are also beneficial in videoconferences, when it may be useful to transmit in advance diagrams which have to be discussed, so that they can be photocopied for each participant. Again, modern faxes have an extra fine resolution for diagrams. Faxes can also provide a rapid record of what was agreed at the meeting, and verbal agreements can in the same way be confirmed in writing. A PC interface enables files to be faxed direct from the computer to another fax, or received directly into the PC, the machine acting as a data modem.

There are as yet few conventions associated with the fax, and this can result in inappropriate use. Faxes may be, and often are, handwritten in a most casual way; if the quality is not good, the result can be unclear or ambiguous. It would, for example, be dangerous to try to proof-read a text produced on thermal paper. Essential information may also be missing—it is not unknown for faxes to arrive in a large organization with no indication of the particular recipient for whom they were intended; there may be no telephone number for reply, when the obvious response is by telephone. Such carelessness is both irritating and time-wasting; it also reflects no credit on the organization that sent the fax in the first place.

This is perhaps the most worrying aspect of the widespread use of faxed material. Companies which would not dream of allowing a business letter to be handwritten will not hesitate to allow a handwritten fax—and yet if it is to be received by another company, all the advantages of a well-presented and professional-looking document are lost. Even more worrying is the likelihood that such a message has not been checked as a letter would have been, and the information may therefore be incorrect, ambiguous or simply ungrammatical. It is not unknown for writers who are angry or in a bad temper to write and send a fax which reveals their state of mind all too clearly; by the time a similar letter had been put on the company letterhead, checked and signed, wiser counsels might have prevailed and it might have been dispatched into the wastepaper basket or the shredder!

The formal and the informal fax

Companies need, therefore, to define their own conventions for the use of the fax. Two uses are widespread, and require differing approaches. Sometimes a fax is used to support a telephone conversation: the writer may want to send some backup details, or to answer a question. Speed of response is essential, and both parties know the circumstances under which the fax is to be sent. Provided that the cover sheet (see Fig. 3.1) is correctly completed, the fax may be informal in style and even handwritten—this is not an ideal, but an acceptance that in practice people are going to send information in this way because of the need to generate an immediate response. Such a message should if possible be followed up by a more formal version, especially if it is likely to be shown to other people.

Most faxes, however, are the equivalent of the business letter: they are formal, correct and binding on the sender. Any document which is to be seen as a representative of its sender's organization should be in this category. Many companies send faxes to other parts of the world, as the easiest, quickest and safest way to convey information to colleagues, clients and customers, and the impact of the message is therefore as important as that of a letter.

Cover sheets

A cover sheet is needed, with the company name, address and logo in a clear, attractive design. A central telephone and fax number may be printed on this sheet, but it is helpful to add the sender's direct line and personal fax number if these will bring a faster response. The sender's name, status or department are needed, and the equivalent information about the recipient, together with any indication of urgency. The number of sheets which make up the fax is helpful, so that the particular document can be easily separated from others which might have arrived about the same time—it is also useful information if there is a problem, for instance if a page has accidentally been omitted. A space in which the subject is identified—as in a memo—allows quick reference to the topic covered. A suggested layout is given in Fig. 3.1; it is assumed that the paper has the company letterhead, with central telephone and fax numbers, at the top.

Facsimile Message

To   Your fax no.

Company   Date

From   No. of pages
                                            (including this page)

Subject Heading

Figure 3.1   Suggested layout for fax cover sheet

This sheet should always be included at the head of a fax; it gives a professional appearance to the document and ensures that critical details are not forgotten.

The rest of the fax has then to live up to this sheet. It will be wordprocessed and checked, and can be relied upon, as is the business letter, as a worthy representative of the company that sends it.

Signing on and off

There are as yet no clear conventions about the greeting or the signing off in faxed material. Most people tend to use the letter formulae of 'Dear Sir or Madam ... Yours faithfully'; 'Dear Mr Jones (Bill) ... Yours sincerely'. This is not strictly necessary, as the relevant information is on the cover sheet, but it is a form which is unlikely to cause offence. Some writers who know the recipient use the less formal 'Bill' at the start and 'Regards' at the end. Both are acceptable provided that they are used appropriately. The general style of writing will tend to follow the greeting, but it is on the whole formal rather than otherwise.

A small point which worries people is the fax equivalent of 'enclosed'; it is probably easiest to use 'follows', as in 'the agenda for the meeting follows this note'.

In the long term, these two ways of using the fax—as an immediate response and as the equivalent of a letter—may move apart and develop their own conventions, or they may acquire restrictions, such as the less formal being used only in-company. For the present, there is a wide diversity in fax presentation; perhaps companies should think through their own usage and the impact of what is sent in their name.

Key points

- Forms and conventions in correspondence change; documents produced by an up-to-date organization should enhance the company image and, where appropriate, follow the corporate style and layout
- The fax is rapid, easy to use and very popular; modern machines allow reasonable confidentiality and can be linked to a computer at either end
- Faxed material represents the company as a letter does, and should look professional and efficient
- Information can also be faxed rapidly and informally to a known recipient; always use a cover sheet
- Decide on the level of formality in faxed messages, and use an appropriate greeting

THE LETTER

In spite of the widespread use of the telephone and the fax, the standard business letter is still a major medium for correspondence. It is comparatively slow and cumbersome in use (the writer is dependent on The Post Office to send it, and it has to be weighed and franked or stamped), but it is reasonably confidential and personal to the sender and recipient. People still send business letters for a range of purposes from job applications to estimates, asking for and sending information, persuading the reader to a particular course of action or reporting what has been accomplished.

Confirmation and reminder

Increasingly, letters are used as written confirmation of what has already been agreed by telephone. There is a certain security in putting details in a letter: both writer and reader know where they stand, and such details as prices, which may be misheard over the telephone, are accurately set down and confirmed. Letters also remind readers, not only of what has been arranged, but of their responsibilities, for instance to pay for work done or to respond to questions or suggestions.

Letter format

Business letters are always formal, following strict conventions about address and layout. As we have said above, punctuation has almost disappeared from addresses, and both sender's and recipient's tend to be blocked at the left-hand side, although a letterhead may have a central position on the page. The date may either follow the sender's address or be on the left-hand side above that of the recipient; it is of course of great importance in a letter, and should always be written in an unambiguous way, that is, with the month written as a word. The most common form of writing the date is probably '3 February 1996'; the old-fashioned style '3.2.96' is too easily read, American-fashion, as the second of March.

Many business letters, sensibly, have a heading which identifies the subject matter, and this is usually placed just above the text, in bold or italic type for clarity. The introductory section of a letter might therefore look like Fig. 3.2 (assuming that there is a letterhead printed on the paper).

---

3 March 1996

Mr J Bloggs
Director, J. Bloggs Limited
Long Barn Lane
Little Melborough
Hampshire LM1 2XB

Dear Mr Bloggs

Company change of address

From Monday 8 April, our company headquarters will be occupying a new site ...

Figure 3.2 Suggested introductory section of a letter

Forms of address

Addressing the reader presents a modern letter writer with particular difficulties. The traditional 'Dear Sir' or 'Dear Sirs' are rarely used, for two reasons: they may be offensive to professional women, and there is a strong and continuing movement towards less formality. In the past, recipients were addressed formally and impersonally unless the writer had actually met them; nowadays, most people would be taken aback to be addressed in this way by someone they had spoken to over the telephone.

On the whole, the formal 'Dear Sir or Madam' should be avoided; it is often better to ring the reader's company and find out the name of the appropriate recipient—this having the additional advantage of allowing the writer to check initials, job title, postcode and other such useful details.

There is a growing feeling that if the writer has not bothered to do this, the letter is unlikely to be of any great significance—which may be an unfair reaction, but is one worth remembering. It is the circular letter, intended for a wide range of readers, which may have to have recourse to an impersonal beginning.

The courtesy of checking the reader's name and position may reveal how he or she wishes to be addressed. In the case of a man, there is a simple choice: he may be called 'Mr' (Dr or whatever), or he may be addressed by his first name. The situation for a female recipient is more complex. Many professional women dislike having their marital status attached to their names, and either wish to use 'Ms' (Dr or whatever), or to be addressed simply by first name and surname ('Dear Susan Smith'). It is a courtesy for such a writer to make her preference clear under her signature: as a general rule, if she prints 'Mrs Susan Smith' under her name, then she presumably has no objection to 'Dear Mrs Smith'; if she simply prints 'Susan Smith', then that is how she likes to be addressed.

There is a general movement towards greater use of the first name, for both men and women; as people so often get to know one another through telephone or fax, they tend to adopt the first name very easily (and certainly at first meeting), and it is used in correspondence far more widely than it was even 10 or 15 years ago.

There are arguments on both sides. The writer may be writing on behalf of the company, addressing the recipient also as representative of a company, and it may reasonably be felt that the address should therefore be formal, even if the two know each other. 'For the attention of Mr W Jones' may be typed above the greeting 'Dear Sir', although this is also less common than it was, and is largely restricted to people who do not know one another. There is also the potential difficulty that it is almost impossible to retreat from informality once it has been established; if 'Dear Bill' suddenly becomes 'Dear Mr Jones', there is obviously a serious rift between them (or perhaps between their companies). It can be safer to remain on a more formal level in a letter, even though the two correspondents use first names over the telephone perfectly happily.

The other argument for greater formality is that, as it is the company which is addressed rather than the individual, the letter may be passed on. The intended recipient may be ill or on leave, and an unknown third party reads the letter. The use of a first name can then make this feel like an intrusion—or, more dangerously, as if there is some extra friendship at work which might prejudice the normal working of the company.

Having said this, in many cases the question of address is smoothly sorted out by the two people concerned; in replying to a letter, it is always worth checking the form used by the previous correspondence, as this is likely to be the safest form to use. One convention especially remains in force: 'Yours faithfully' is the only acceptable ending to a letter which begins without a personal name (that is, with Dear Sir or equivalent), while 'Yours sincerely' is the general ending to business letters which have begun with the use of a personal name.

Style and usage in letters

Each line of a letter is nowadays blocked to the left, with a double space between paragraphs. Punctuation within the text is normal, with the constraints about sentence length and accuracy which were described in Chapter 2, Writing Techniques (see page 9). However, writing letters sometimes brings out oddities of style which are worthy of comment.

As letters are formal documents, the writing style (and of course the checking) is also formal. Formality, however, can easily become pomposity, which should be avoided. The use of abbreviations such as 'yours of the first inst.' and so on has almost died out, but the expression 'your esteemed correspondence' was noticed recently. Such writing can become long-winded, as in the amazing but real-life example:

We offer our express apologies and hope that the situation has been explained and rectified and offer you our future best intentions ...

This presumably means 'We apologize. We hope that the problem has been put right. We'll try to do better in the future.', which in itself sounds clear enough but a bit over the top. 'We apologize and will try to ensure that such a problem does not arise again' seems to be quite as much as is needed!

There is a strange reluctance on the part of letter writers to use the words 'I', 'me' and 'you', even when they are obviously meant. This can result in the awkwardness of 'Please contact the undersigned' instead of 'Please contact me [us].' It can produce even more tortuous writing, as in 'We ourselves will be happy to reimburse yourselves'. Such problems occur particularly at the beginning and the end of letters: 'Further to your letter of 5 May' is not a sentence any more than 'With reference to your letter' is, and yet both are not uncommon; 'Thanking you in advance' is another non-sentence which sounds unwieldy (why not just 'Thank you'?). Trying to say the right thing without thinking it through has also produced accidental humour, as in 'We look forward to hearing from you with interest'; unfortunately, turning this round to 'We look forward with interest to hearing from you' makes it sound like a challenge—'get out of this if you can!'

As in most technical writing, the simple approach is the best. Many letters can start with a straightforward courtesy, such as 'Thank you for your letter of 5 May' and conclude with 'We look forward to hearing from you'. If the situation is more complicated, it is usually wise to start with the good news in order to build a good relationship with the reader and then to move on to the problems ('this has been completed satisfactorily ... However, ... ') Bad news is often indicated unconsciously by a rambling start: 'I am writing this letter in order to inform you that ... ' is unlikely to precede the offer of a job!

Letters are usually quite short, mostly not more than a page, or two at the most. If they become longer than that, especially if they contain technical details, it is more helpful to the reader to divide the material into a short covering letter and a report. There is a clearer structure to a report, and the headings and subheadings make identification of information much easier than it is in a letter. Even a two-page report will be easier to use, and the message is likely to reach more people and be more influential than if it is presented as a continuous narrative. It is possible to use more than one heading in a letter, or to list points—but at this stage it is just as easy and more satisfactory to turn the letter into a short report.

Key points

- Business letters are always formal in style and layout
- Choose an appropriate form of address, checking if necessary how the recipient wishes to be named
- Do not write in a pompous, long-winded style—clarity and courtesy are essential
- If you have technical material to communicate, consider the advantages of a covering letter and a short report

ELECTRONIC MAIL (E-MAIL)

Alongside the fax, e-mail has developed as a comparatively fast means of transmitting information across a computer network. In some ways it resembles the postal system (as its name suggests), as the mail goes from the sender to a storage and sorting centre, and then passes on to the receiver's computer screen when it is open to accept the message.

In theory the system should work efficiently, and often it does, but there are potential and sometimes actual difficulties. E-mail is not always as fast as the fax, and it is less easy to find out whether the message has got through or not (faxes work only when the route between sender's machine and recipient's machine is open and clear, and inform the sender if this is not so). E-mail can be read only when the recipient is prepared to receive it and uses the appropriate password; it has been known for mail to build up at the receiver's end and remain unread for a considerable time.

E-mail can also get lost in the system: messages may unaccountably arrive late (or not at all?), and its security is questionable.

In spite of these difficulties, e-mail has to a certain extent overtaken the memo as a means of passing information within an organization. It allows technical material to be sent clearly and quickly, and is capable of producing a rapid response. E-mail is on the whole easier to absorb than a message on an answerphone, which is sometimes rushed or unclear. In addition, it is a useful method of sharing documents: information can be sent and received world-wide—a report, for instance, could be transmitted via e-mail to readers in many parts of the world. Educational material can travel direct to students from a tutor who might choose to send it from the comfort of home; the same tutor might also use the system to check details of research with specialists at the other side of the world.

There is obviously a good deal of potential in an e-mail network, and yet there is also resistance to using it as a world-wide medium. Lack of security is part of the difficulty; another is the impossibility of getting the message through if the recipient does not choose to complete the link. It has also been found to generate paperwork rather than to make it redundant: messages tend to be printed out in order to be kept/considered/acted upon/taken to a meeting, and so on.

E-mail is also undiscriminating. It accumulates messages of varying degrees of importance or urgency, and the recipient has to look at them in order to decide whether they are worth reading or not. Perhaps just because it can be a friendly means of communication, it seems to inspire junk mail, and irritation at a collection of time-wasting trivia can dissuade people from looking to see if there are important messages awaiting attention.

There seems to be little control of e-mail at present. Companies may feel that they lose the sense of corporate identity which a fax can provide, and they may need to insure against litigation arising from inaccurate, incomplete or unlawful information transmitted in this way. In the future, many of these disadvantages may be overcome, but for the present, organizations often prefer to restrict e-mail to internal use and choose the fax as the major means of communication with the rest of the world.

Key points

- E-mail is a quick, easy way of passing information or getting a fast response; it depends on the person at the receiving end choosing to 'collect' it
- Documents can be shared world-wide through an e-mail network
- E-mail is not very secure, and is undiscriminating—it can become a repository of junk mail

THE MEMO

In spite of modern technology, some forms of correspondence have remained almost unchanged for generations: the memo is perhaps the prime example. Originally, memos were seen as a quick, informal way of sending information to colleagues; they rarely travelled outside the building in which they were generated and never outside the company, and were used to pass messages to small numbers of people, usually of about the same level on the hierarchy.

On the whole, this is how memos are still used today. They may sometimes turn into a kind of notice, pinned on a notice board for passers-by to see and perhaps read, although they are not designed for this purpose and rarely attract much attention. A reminder about attending someone's leaving party represents a typical use of this sort. Memos still tend to be used between colleagues and on the whole their circulation is small; as a result, they are the least formal of the traditional means of correspondence.

By convention, memos have four headings, usually arranged in pairs at opposite sides of the page. These are 'To', 'From', 'Date' and 'Subject' (or sometimes 'Reference'). Such headings are usually printed on a memo pad, which may duplicate the information automatically. This suggests that a memo, unlike a letter, may be handwritten, although this is rarer than in the past, and the headings are more likely to be formatted on the computer and called up as needed. The company may still require a copy for filing.

Memos are usually short, generally less than a page, and should deal with one subject only. The length of the circulation list can be a nuisance—it occasionally turns into the longest piece of writing on the page. In this case it may be easier to print the list below the message and to highlight the individual recipient's name. The style may be less formal—though not less correct—than that of a letter or a fax. Abbreviations such as 'can't' or 'it's' are acceptable, as are 'I' and 'you'. There is no need for a greeting at the start, as the recipient's name is shown, and indeed no need for a signature, as the sender's name is also given. Nevertheless, most companies insist that memos are signed (though sometimes a first name or initials are sufficient).

Memo envelopes, which can be reused, are common, and sometimes the memo is simply folded over and stapled. It is not worth putting a memo into a proper envelope, and for this reason a memo should not contain any material which is confidential or restricted in any way.

The telephone and e-mail together may supersede the memo, but for the time being memos continue to fulfil their traditional role in many organizations. They may add to the paperwork, but they represent a written record—and as such, they need at least a little thought and planning. It may even be worth drawing a quick spider diagram (see page 13) in order to structure the information; the details should at least be ordered in a way which is helpful to the reader. The following example shows a real-life memo which is almost incomprehensible. The writer obviously failed to consider its impact on the reader:

The customer defect form is in three sections. The pink section, which copies through to the blue and white sections, is filled in by the site representative or customer and is kept by the originator. The blue and white sections are returned to

QA together if the repair is carried out on site, but if not, the white section is sent with the equipment and the blue section to QA. QA log the defect and keep the blue if it is attached to the white, the white being sent to Projects for comment and returned to QA. If only the blue is returned, it also is sent to Projects for comment; they enter their recommendations and return the form to QA. The completed blue section is returned to QA by the investigators and the completed white from the repair department ...

This unhelpful information can be made to fall into three parts, after the normal to, from, date and subject headings:

1 The top (pink) copy of the customer defect form is filled in by the site representative or customer, and is kept by the originator. Two further copies, one blue and one white, are made automatically when the top form is completed.
2 **If the repair is carried out on site**, both blue and white forms are returned to QA. QA logs the defect and retains the blue form. The white is sent to Projects for comment, and returned to QA.
3 **If the repair is not carried out on site**, the white form is sent with the equipment and is returned to QA on completion of the repair. The blue form is sent to Projects for comment and recommendations, and then returned to QA.

The effect of highlighting the two key messages ('If the repair is carried out on site' and 'If the repair is not carried out on site') is to carry the reader straight from the introductory details to the relevant information. Readers can avoid cluttering up their minds with irrelevancies, and the message is therefore both shorter and easier to remember. Memos should always be short; if they also convey the right information simply and clearly, they may continue to play their part in company communication in spite of the alternatives currently available.

Key points

- Memos are slow and cumbersome, but they are still effective in allowing colleagues to exchange information
- A memo is short and can be informal; company policy may dictate that it is signed, and a copy kept
- All forms of correspondence should be planned, organized and checked

4
Written Text

Reports • Specifications • Documentation for meetings • Procedures and instructions

REPORTS

> Why, a four year old child could understand this report. Run out and find me a four year old child. I can't make head or tail of it.
>
> *(with apologies to Groucho Marx)*

Many report readers will sympathize with this sentiment: reports are often unattractive, unstructured and verbose, and undermine the reader's attempts to extract information rapidly and efficiently.

Readers and objectives

Reports—except for the very shortest—are rarely read from cover to cover like a detective novel. Readers, or more accurately users, want to extract specific material: it is needed urgently, or it is particularly important, or they have to go to a meeting on the subject, or perhaps they simply need the answer to a question. If they can find the information they need quickly and accurately, they feel that the report is a good one; if they cannot, they may not bother to return to it at all.

Technical people write reports for many reasons, mostly to do with persuasion. They describe the work achieved so far and look to the next stage, ask for more time or equipment, account for the time they have spent in visiting another organization, seek agreement for a course of action or perhaps simply describe a complex piece of work so that others can understand it (and perhaps give it further support).

There are two sets of objectives, the reader's and the writer's, and both should be analysed in as much detail as possible at the very start of the report-writing process. Why is the report needed? Who is going to use it—one person, a clearly defined group of people, readers of differing interests and perhaps expertise? What are they hoping to get out of the document? Are they looking for specific information or ideas? The question of the readers' objectives is closely linked to the readers themselves, their background, prior knowledge and experience. All of this information should be available to the writer, and all of it is important. It may also be useful to consider what cannot be ascertained: will the obvious reader in the company receiving the report pass it on to other, unknown readers? Will unexpected, perhaps financial, decisions be made on the basis of the technical data?

Such questions are essential as the writer starts to prepare the report. The document must be focused as far as possible on the specific needs of the readers, and the level of technical detail and terminology must be appropriately chosen.

The writer also has objectives, not just those suggested above, but others which are less often identified. The report must show the writer in a good light, as a competent professional; it must also reflect well on the writer's company, and maybe even on the profession as a whole. Reports may be the only end product of the organization, and the impact they make on the client could be the most important 'marketing' undertaken.

Usually, the two sets of objectives coincide—the writer wants to 'sell' what the reader wants to 'buy'—but occasionally there may be a conflict. The writer may want to put forward a suggested course of action, while the reader may have already decided to do something else and be reading the report primarily to find fault with it. There is no easy answer to this, but if the report makes a good impression, the writer may be consulted again in the future.

Collecting information

In the light of the readers and the objectives, appropriate information must be gathered. The sources are many: the writer's own work and that of colleagues, previous reports and other publications, surveys, discussions and so on. All relevant leads should be followed up, so that the information presented is as complete as possible, although in practice the major difficulty is much less one of what to include than of what to leave out.

In this, the objectives already identified are the greatest help. Is this information relevant to my objectives, and if not, why am I trying to include it in my report? There may of course be a good answer: for self-protection, for instance if the report might be used as evidence in legal proceedings. Such an answer is perfectly valid; however, 'because it took a great deal of trouble to get this information' is not. Much of the excess in reports is there for such reasons—the writer has a vested interest in including it, it seemed interesting at the time, it was difficult to get hold of, it was much easier to obtain than other information which had to be left out. If the information is not relevant to the objectives and if there is no other *good* reason for including it, leave it out.

Making—and checking—assumptions

One of the most difficult aspects of this early stage of planning the report is to assess what can be taken for granted. Assumptions are always a problem. We all tend to assume that what we know is common knowledge, and the result may be to omit important areas of information or explanation, and to bewilder the reader. The opposite is just as bad—to over-explain what any imaginable reader is certain to understand, so seeming to patronize. (A glossary is useful in overcoming the difficulty—see page 55.) Some assumptions must be made, for instance that writer and reader share a common language or that a translation is available, but to assume too much or too little familiarity with the subject can lead to a report whose message or whose reasoning is unclear. Identifying the reader and the objectives is an essential step in deciding the assumptions that can be made, but it is not the end of the story. We must make allowance for other readers—colleagues of our known reader, for instance—and for the fact that a report may be used out of context, when new circumstances suggest that its subject is again topical. The original writer and reader may not be available to explain what was written, but it must still make sense.

Before the report is finally printed and sent to the reader, it is also worth going back to those original decisions and asking the questions again. Have I now given the reader enough background, enough detail, enough explanation, so that the objectives of the report (mine, anyway) can be fulfilled? If not, where is the problem, and can I do anything further to improve the situation? Only when we are sure that we have collected all the relevant information and produced it in the most appropriate way can we feel that the report is nearing completion.

Main text and appendices

As soon as the writer has identified readers and objectives, it is time to start thinking about structure. An outline contents list may already be a possibility (see page 14 for ways of achieving this), and as the information is gathered, the structure will become clearer and more detailed. One of the first decisions will be about what is to go into the main text and what properly belongs to the appendices.

Appendices are the report writer's friends. At the back of the report can go supporting evidence, detailed mathematical calculations, data of interest to only a small minority of readers, possibilities which were finally excluded but which might be referred to in the light of future developments—all the information which might otherwise clutter up the main text of the report and make it harder to read and use. As the report material is collected, it is useful to think of it as being in one of three categories: obviously essential and so in the main text; backup and so in the appendices; probably irrelevant and so to be kept on one side just in case it is needed and eventually filed away for possible future use in a different document.

Report structures

Reports should have a clear and logical format. They are different from essays or articles in that their logic must not only be there but must be seen to be there, in the use of headings and so in the contents list in which all the headings should be included. A good test of the quality of a report is to look at the list of headings and to see how far this gives an indication of the structure, what is included and how the thinking develops. If it does this, the report is likely to be a good one; if it does not, then the reader is disheartened from the start.

Many companies have an established format for reports and other documents, available through the computer network. This often gives instructions about the choice and format of headings, and about the detail of presentation, such as title pages and margin sizes. Obviously, staff are expected to follow these guidelines, although it is always wise to leave some flexibility for the document which does not comfortably fit the pattern. Nevertheless, it is very helpful to be able to call on such support, especially to those new to the organization.

It is also possible to use the format of a previous report, especially on a similar or related subject, and to adapt it as required. Again, this starts the writer off in a positive way, allowing some decisions to be made quickly and easily.

Where neither company format nor a previous report is available, the report writer is not left totally without help. There is a basic structure to which nearly all reports conform, and which will act as a starting point:

Introduction and Scope
Evidence (Findings)
Conclusions on the basis of the evidence
Recommendations, if asked for, on the basis of the conclusions
Appendix material

This is a primitive arrangement, and not nearly detailed enough for the final document, but it is a pattern to start with, especially if the writer can immediately see some of the main subdivisions of the evidence.

Many technical reports follow this basic pattern, often in the more detailed form suggested below; it is, however, only a suggestion, as report formats vary with the type of report, the subject, and, of course, company house style. The major sections will be discussed in the order in which they are likely to appear.

Title page
A title page gives a report status; it should contain the 'administrative' details, for instance:

Short title
Author's name and department/company
Date
Client's name and company details, if appropriate
Reference number
Issue number and date, if the report has been reissued
Signatures of authorizing officers, as appropriate
Classification such as Confidential, also on every page if appropriate

Summary
The summary is potentially the most influential section of a report. Some people will read only the summary (see below), and almost everybody who reads the report will start with the summary. It must make sense by itself, without reference to the rest of the text, and should contain a short statement of the problem, an indication of the essentials of the evidence, and, in brief, the main conclusions and recommendations. As its name suggests, the summary should be short—for a report up to 30 pages long, 120 words should be sufficient. A rather longer version, usually known as the executive summary, is sometimes called for; it should be treated with great caution, as it often loses the weighting of the full text and includes detail which is not appropriate for a summary.

Abstract
An abstract guides potential readers to the subject matter of the report, often identified by three or four key words. Abstracts are unusual in industrial reports, but are sometimes found in the reports of very large organizations.

Contents List
As few reports contain an index, the contents list has to take its place in guiding the reader through the text. It should show all the section numbers and headings, and page numbers unless the report is to be transmitted electronically. A list of diagrams may also be included.

Introduction
The introduction tells the reader why the report has been written, and explains the background; it may contain any particular terms or symbols which the reader may not know (but see Glossary), or any other general

information which will help the understanding of the technical data which follow.

Scope
Recently, an increasing number of reports have included a scope section, which shows the coverage of the document, with the depth/extent of the discussion. It may also show what is excluded.

Glossary
There is no clear convention about where to place the glossary: it may be, as here, before the main body of evidence, or at the end of the report. In a technical document, it is very important as it shows the meaning of any terms, signs, symbols and abbreviations with which readers may not be familiar. Part of its value is in helping readers who have expertise that is different from or less than that of other readers, while the knowledgeable can simply ignore the page, without feeling irritated by constant explanation of what they already know.

Evidence (Findings)
The evidence or findings should be presented as objectively as possible (but in choosing what to include and in what order, the writer is inevitably showing a point of view). This area of the report will be divided into numbered sections and subsections as the subject matter suggests (see below). Diagrams (see page 29) may be included here or in an appendix.

Conclusions
The implications of the evidence presented are given in the conclusions. No new material should be introduced at this point; the writer assesses the evidence which has been produced.

Recommendations
The recommendations section, if asked for, puts forward one or more possible courses of action to be taken as a result of the conclusions which have been drawn. The writer may be subjective in giving an opinion, indeed, his or her professional credibility may rest on proposing a balanced, workable and effective solution to the problem which gave rise to the report. The recommendations will have been given in brief in the summary; here they are expanded and discussed in detail.

Appendices
Detailed evidence, of interest to a minority readership only, may be given in an appendix or series of appendices (see page 10). Diagrams which are purely supplementary or which will not fit the text may also be put into the appendix.

References
Any books, journals or other reports which have been used or quoted should be listed with full bibliographical details.

Acknowledgements
It is unusual to find acknowledgements in an industrial report, but where they are needed they may be at either the beginning or end of the document. By convention, acknowledgement is not made of the help of colleagues, but only of outside individuals or organizations who have given help, advice or information.

Annexes
Separate documents such as previous reports or specifications may be bound at the end of a report if this will help the reader. Such annexed material has its own page and section numbering scheme, and so is clearly not an integral part of the main document. Both appendices and annexes should be included in the contents list.

One or two of these sections require further comment.

The **summary**, as has been suggested, is likely to be the most influential section of the whole document. Virtually everybody who reads the report will start by reading the summary, and if they find the subject matter difficult, perhaps because of different expertise, they may rely heavily on the summary to clarify the general sense of the report before they move to the technical detail. Other readers, needing a quick overview before settling down to deal with the detail, may read the summary several hours or even days before reading any other section. Yet others, who have read the whole report, will reread the summary as a reminder of the main points, perhaps before going to a meeting on the subject.

Even more critically, some readers will read only the summary. Perhaps their work is peripheral to the subject, or they simply want to know in general terms what is going on: the summary supplies exactly what they need. As reports move through companies, they tend to go slowly up the management hierarchy, and as they do this, less and less of the report is read. The most senior reader, who may well make the necessary decisions, will probably read only the summary, perhaps passing on the document to other people for technical analysis.

The importance of the summary is clear. On all these counts, it must be effectively written (generally as continuous prose rather than as a list), concise, and placed at the beginning of the text, which is where it will be needed. Most reports benefit hugely from having a summary, and its absence often represents a wasted opportunity.

The **title page** is the first part of the report to make an impact on the reader, and it is therefore important that it is well-designed and contains only the appropriate information—the inclusion of a summary or contents list clutters the page and damages its effect. Many organizations have standardized title pages, to reinforce the company image.

Technical reports usually have to be countersigned by senior managers before they can be issued, and such signatures are generally found on the title page. As long as there are only two or three signatures, this is reasonable, but we have seen instances where the process is carried to excess: five or six signatures look confusing, and in our experience such bureaucracy tends to undermine itself, as report writers will try to find a way round the system in order to circulate the information more quickly.

The title page should always contain at least one date; in the case of a visit report, for instance, there will be two—the date of the visit and the date of the report. All dates should be written in an unambiguous way. The American format of month/day/year has been disseminated by the sale of watches with the date in this order and by the standard American date order used by computer software systems, and as a result the traditional British way of writing dates, for instance 1/12/96, has become unclear. Most technical reports use only the month and year, but if a more precise date is needed, it should be written as 1 December 1996, which is unambiguous.

The **conclusions** and **recommendations** of even a very short report should be kept separate. In these two sections, the writer is being subjective in two different ways. The conclusions form an assessment of the implications of the evidence, while the recommendations suggest a course of action to be taken as a result of the conclusions. If the two are merged, the recommendations will tend to get lost, and some may be overlooked. The separate headings guide the reader to the appropriate stage of the report.

The relationship between these sections of a report may be illustrated by the simple, real-life example of a garden wall belonging to one of the authors. The main sections are shown in outline only.

Introduction
Report prepared by ... after inspection on ...
Problem ... the wall is three feet high on one side and twenty feet high on the other. It is cracked and appears to be leaning towards the (lower) garden of the neighbouring house.
Constraints ... the wall is 120 years old, and is the boundary wall of a conservation area. It is also listed, so that its appearance must remain the same after any work is carried out. There is restricted access alongside both (detached) houses.

Evidence
The wall has been inspected, and is described in detail. The extent of cracking and of the lean (8 degrees) is shown, with plans and diagrams. Situation, ground conditions, lack of drainage points, etc. are all given.

Conclusions
The wall is leaning dangerously for reasons which are now clear, such as the

absence of drainage points at the base of the wall. Action should be taken as soon as possible.

Recommendations
There are two or three possible courses of action: the recommended solution is a trench dug at the upper side of the wall and filled with concrete; ties to the wall at intervals would hold the wall back and prevent further cracking/leaning. The wall should be checked at intervals in the future.

Appendices
A Technical details of the proposed solution.
B Some detail of the 'second best' solution, in case the first choice proves impossible.
C Estimates of cost, time, etc. involved in carrying out the recommendation.

Summary (prepared last, but to go at the start)
The wall is dangerous ... immediate action is necessary ... the proposed solution in outline.

Headings and numbering systems

We said earlier that the logical pattern of any report should be revealed through its headings; any of the sections discussed above might be divided into subsections with appropriate subheadings. A heading serves more than one purpose in a report: it reveals the subject matter of the section or subsection that follows; it gives emphasis to that topic (major emphasis in the case of a major heading, lesser but still significant emphasis in the case of a subheading), and, in company with the other headings in the contents page, it shows the logical structure of the whole report and, to a certain extent, shows what is omitted from the subject coverage.

For all these reasons, every heading must be as precise as possible. Vague headings, such as *Miscellaneous* or *General* are unhelpful, as the reader's attention is not drawn to any particular information—as a result, the section may well be ignored. The more exactly the heading identifies the content, the more secure the reader feels in extracting the specific information needed. A word such as 'infrastructure', for instance, is very wide; if it is used in a heading, the reader might well feel cheated, in perhaps finding information about the mains services but not about the roads. The words chosen in a heading should be narrowed down to fit the precise ideas which follow.

Headings in reports are numbered, so that the hierarchy of information is clarified and sections and subsections can easily be identified and extracted. There are various systems available, although for technical material two systems predominate: paragraph numbering and decimal notation.

In many ways they are alike, as both use arabic numerals in a hierarchical fashion, although decimal notation carries this logic through every level of the report. Paragraph numbering varies, but perhaps the most common form has numbered headings and then under each heading the numbered paragraphs that give the information. This has the advantage of easy reference, for instance in a telephone discussion; its disadvantage is that the logic breaks down into sequential numbering; paragraph 64 is so numbered because it comes after 63 and before 65—there is no obvious logical pattern to help the reader.

The other common problem with paragraph numbering is that it tends to result in very long paragraphs; writers are unwilling to start a new number (paragraph) if they are still discussing the same topic, and so allow the paragraph to become unwieldy and difficult to read (see page 27).

In decimal notation paragraphing is irrelevant, as it is the headings that are numbered—indeed, the principle of decimal notation is that numbers and headings always go together. Major headings are given single arabic numerals; when subdivisions are needed, a decimal point is used, and the numbering starts again (1.1, 1.2, 1.3, etc.). If further subdivisions are required, the process is repeated (1.1.1, 1.1.2, 1.1.3, etc.), and yet another subdivision is available if the material is sufficiently complex (1.1.1.1, 1.1.1.2, 1.1.1.3, etc.). Four levels should be enough: if the writer feels the need to use more, it is likely that the material has not been well structured.

The format of the headings should reinforce the numbering, so that the major headings all look the same and clearly are of major significance, and lesser headings show their importance/subordination with an appropriate format. In spite of its name, the system is not truly decimal, as the numbers do not have to stop at 9—it is perfectly possible to have subsection 2.4.16, if the information dictates this. The great strengths of this system are that it is clearly logical, easy to use, widely recognized and governed in practice by the needs of the material itself.

The following example shows the system as it is used in an extract from the *RICS Homebuyers' Survey and Valuation Report*. It is reproduced by permission of the Royal Institution of Chartered Surveyors which owns the copyright.

3   Common Services
3.1 Water and heating
3.2 Lifts
3.3 Security system
3.4 Fire escapes

As can be seen, it is easy to extract a particular interest, for example lifts, and to use that information without either struggling to find it or being unsure where it ends—clearly it starts at the 3.2 heading and is completed by the 3.3 heading.

Within the numbering system, information may need to be listed (see page 28). It is better in this case to leave the decimal system intact and to number the items in the list by themselves, as (1), (2), (3), etc., or (a), (b), (c), etc., as preferred. On the whole, it is wiser to use the arabic numerals, and to leave letters for the identification of appendices. If Appendix A, Appendix B, etc. are used, there is always the possibility of subdividing an appendix if necessary, so that Appendix A1.3 would show the third subsection of the first section in Appendix A.

Style in reports

Good style in reports is much the same as good style in any other form of technical writing, as discussed in detail in Chapter 2. It is, however, important that a report remains a formal document; it is as unnerving for readers to feel that they are reading a conversation in print as it is to an audience to feel that it is listening to a book talking. The English language distinguishes between the spoken and the written language, and also between a comparatively informal piece of writing, such as a note to a friend, and more formal 'business' writing, such as reports.

A modern report may not be written in the passive form (see page 16), and its writer may be able to use the first person plural ('we'), but it will never become so personal that it includes either 'I' or a direct address to the reader as 'you'. Needless to say, there is no place for slang or abbreviations of the 'can't' or 'won't' variety, but at the same time the report must not become pompous. Too much technical writing is long-winded and over-complicated; a simple, concise, direct form of writing appeals to the busy reader and is likely to encourage further reading.

One further aspect of the writing of a report is important and often misunderstood. It is sometimes said that the evidence in a report should be given totally straight, in a completely unbiased way, and that the writer's opinion should be reserved for the conclusions and recommendations. In principle, of course, this is true. Nevertheless, the writer makes choices from the start of the report preparation, and in deciding what to include and what to leave out—and in the ordering of information—he or she is necessarily showing personal prejudice. No information is ever given without some element of personal involvement. It is, however, important that the writer should be as objective *as possible* in presenting the evidence, or else the reader will become suspicious and start to distrust what is written. The writer has therefore to be very careful, in choosing words, to be fair to the information and to the reader, and to present the material in as unprejudiced a way as can be managed, reserving opinions for the appropriate sections.

Presentation of reports

A report should be as professional and as impressive in appearance as it is in its technical data. If the information is incomplete or inappropriate, the report will not be successful; some reports, however, are ignored because they look unattractive or difficult to read. Every report, as we have said, is an ambassador for its writer and maybe for the company, and it is a waste of everyone's effort if it is ignored.

Bindings and covers

The first part of a report to come to the reader's attention is the cover. Does it fulfil its task of holding the report together and allowing it to be identified quickly? Many reports fail even at this stage, by having too many pages in an inadequate binding, so that the spine breaks and some pages become loose (and therefore are lost). The opposite case, when there are only a few pages in too thick a spiral binding, also causes difficulties, for instance when copies of the document are together on a shelf and the spiral bindings become interlocked, so that it is almost impossible to separate the reports. The form of binding must be compatible with the length of the document and also with the way in which it is to be used. If, for example, there are many detailed diagrams which must be studied, the report must lie flat on the table, without snapping shut every time the reader moves away a restraining hand. The text must fit the binding: a common problem with student reports is that the text goes too far to the left-hand side of the page, so that the popular slide-bar binding covers the first word on every line, and the whole report has to be dismantled in order to be read.

The cover of the report should be appropriate for its importance. On the whole, short in-company reports can have spiral bindings and a simple, standardized cover with a 'window' showing the author and title. Identifying the report is more important than impressing the reader. If the report is to go outside the company, the impression it makes is very important; the report may be bound in the company colour, with perhaps the company name and logo added to the other information shown. It is often worth getting professional advice from a designer, as such a cover can become standard for all the company's documentation and will rapidly become associated with the organization in the minds of the readers.

Increasingly, reports are stored electronically rather than in paper form, and this will in time become the norm. There are obvious advantages in terms of storage space and ease of access. Nevertheless, it is important to use an easily-identified file reference for the report and to have backup for all material so stored, and to ensure that it can always be retrieved—systems can become obsolete, and if the disk can no longer be read, the material is lost. In practice, it is likely that it *can* be retrieved by an expert, but time and money will be wasted.

Layout of title pages

After seeing the cover, the reader will probably look at the title page and get a first impression of the report itself by doing so. As with the cover, it is wise to seek professional help from a designer, to be sure that the title page is both professional and attractive in appearance. It should then be standardized throughout the organization. However, in too many cases the layout of the title page is left to the individual writer, with some strange results. After all, technical people are not usually trained in design or typography, and it is not surprising that they may not, for instance, realize that a block of upper case letters is harder to read than a natural mixture of upper and lower case letters, or that a wild mixture of fonts can look peculiar and oddly amateur on the page. Above all, the title page should not look cluttered, and its most important information—author, title and date—should be immediately accessible.

Layout of text

Each individual page of a report should be designed for the ease and encouragement of the reader. Space is essential—there are few things more discouraging than a page of crowded text, so that the reader can see no point at which to pause and assimilate the data so far. The sections and subsections of a report help to create an impression of a text which is broken into manageable chunks, and the headings—which should always be on a separate line from the text—provide a kind of introduction to what follows. Nevertheless, the writer should be aware of the need for good margin space and space between paragraphs; these are often standardized on the company's computer system.

The font chosen for the text is largely a matter of taste (or, again, company policy). It should be as clear and uncluttered as possible and should not distort the letters. A sanserif font is usually recommended for technical material, not smaller than 10 point. Small print is difficult to read, especially for older (more senior?) readers, and for this reason the temptation to reduce the text on the photocopier should be resisted.

One font is usually sufficient, especially if bold type is used for the headings and italic for particular messages such as warnings, notes to diagrams or key sentences. The length of line is also a factor in the readability of text; if the number of 'units' (a key stroke, whether letter, punctuation or space) exceeds 80, the line is too long for easy reading (and the font is probably too small). The normal line spacing on the wordprocessor is usually acceptable, but it may have to be adapted if there are many superscript or subscript numbers in the text; double spacing is harder to read, although often required for marking purposes, for instance in dissertations.

The overall impact of the page and of the report is important, and we would stress the advantage of using a professional designer. If there is no opportunity to do this, the writer should at least ask a colleague to look at a page of text to make sure that the headings stand out clearly and the text looks accessible. The writer should also try to take an objective view of the whole document, bearing in mind that it must win the goodwill of its readers in order to be successful.

Key points

- A report should be organized so that readers can extract information accurately and rapidly
- Identify the reader or readers—but do not forget that others may also need to use the report
- Identify both your own objectives and those of the reader before you start
- Collect as much information as possible, testing it against the objectives to see if it is relevant
- Plan the structure of the report before starting to write
- Choose a format which is logical, easy to use and, if possible, widely recognized
- The summary is often the most widely read and influential section of the report
- Headings should reflect the information below, as accurately and precisely as possible
- Choose a clear and logical numbering system
- Present your evidence as objectively as possible; your conclusions and recommendations will reveal your assessment of the facts
- A report should look as impressive as its contents; it acts as an ambassador for its company
- The layout of the text should attract the reader and encourage the reading
- A well-written, well-presented report wins the goodwill of its readers

SPECIFICATIONS

The word 'specifications' covers a wide range of documents: material specifications describe the materials or components used in manufacturing a product; test specifications describe methods of testing with the appropriate criteria for passing or failing the tests; other functions such as installing, operating and maintaining machinery or systems are each detailed in appropriate specifications.

Essentially, a specification sets out requirements; it has therefore to be written in a clear, unambiguous way which the intended reader can follow and apply. An error, of fact or printing, may have disastrous consequences legally, or morally, in terms of an accident.

As with any other technical document, the specifier must start by identifying the user and the operation to be described. Who will carry out the work, and what is the level of that person's knowledge and responsibility? Is the specification to be written for users of the product, or for those who will manufacture, install or maintain it? Only in the light of this knowledge can the writer decide at what level to give the technical data: it is obviously dangerous to bewilder the reader with too much technical content or language, while a detailed explanation of what any conceivable recipient of the information will have known since schooldays causes frustration and annoyance—and may result in a skimming of what should be read in detail. Once this level has been established, it must be adhered to throughout the document.

Format of specifications

Specifications, like reports, have a structure, which must be planned before the writing can begin. A spider diagram (see page 13) or similar planning technique is useful here, especially to help the specifier organize what is introductory and what relates to aspects of the main content, for instance, in a design specification the materials, appearance, maintainability and so on. Again as in a report, the scope of the specification will need to be clarified and a contents list sketched, so that the writer has a clear logical pattern to write to. A suggested order of information is given below.

Title page including title, date, originator, reference number, issue number if appropriate

Contents list giving sections, headings, page numbers

Introduction giving general information about the circumstances under which the specification is to be used, its objectives, and any necessary background

Scope informing the user about the range and depth of the coverage and also about limitations, exclusions, etc.

References both normative and informative (see below) as appropriate

Definitions of specified names, terms, etc. which retain their defined form throughout the document (see below)

Glossary setting out definitions of technical terms and abbreviations which the user might not be familiar with, or which are used in a highly specific way (such terms and abbreviations should if possible conform to British or International Standard usage)

Main body of the specification including the materials, methods, etc. described in detail, in appropriate sections

Appendices giving supporting information or additional details which might be helpful to the user

Obviously, specifications will vary in detail depending on their subject and use, but many will have a structure similar to this. Some sections, such as the contents list, glossary and appendices, will be much the same as their equivalents in a technical report, but others need especial comment.

References

References have a particular importance in a specification: it is likely that the specifier will need to quote provisions from other documents, as few specifications will contain every important detail in themselves: cross-referencing to other specifications for some of the provisions is therefore common. Such references are known as *normative references*, and they are listed near the beginning of the document so that readers will know what other material is needed in order to supplement the specification which they are using. It may also be important to define the relationship of such references to the specification itself, in some such paragraph as the following:

> References in brackets in clause headings in this Specification are to clauses or subclauses of BS 8110 'Structural use of concrete'. Unless modified by this Specification, concrete work shall comply with the requirements and recommendations of these clauses or subclauses. In cases of conflict, this Specification takes precedence over BS 8110.

Normative references are often followed by *informative references*, which list other publications which will be helpful in providing guidance or related information. Such references will usually be given with full bibliographical details, as in the case of technical reports (see page 56).

Definitions

Some names or terms have a specific meaning within the document, and must be used consistently throughout. It is a convention of specification writing that such terms have initial capital letters, so that the specific is easily distinguished from more general usage. For instance, Client will refer to the client for whom the document is being prepared rather than to any other client; Contractor will have a similar precise reference. Such words as Design Drawings, Works or Plant will have initial capital letters to show their specific meaning. In effect, 'the Contractor' is the particular company named in the document, while 'the contractor' is any *other* contractor who might be mentioned. It is therefore essential that such use of capital letters is consistent throughout the specification.

Other definitions might be needed, such as definitions of time ('a clear day' is taken to mean 24 hours, starting at midnight), place or cost. Part of the checking system of a specification is to ensure that all necessary definitions are included and that they are used consistently by everyone involved in producing the document.

Main body of the text

A specification, again like a report, is divided into sections with headings. In the nature of the document, it is likely to have far more subdivisions, perhaps getting to the fifth or sixth level of decimal notation and even then needing a, b, c, etc. at a further level. Individual paragraphs may also be numbered within the section. The contents list of a general concrete specification, for example, will show a section with its major divisions as follows:

2.0 MATERIALS
2.1 **Cement**
2.2 **Aggregates**
2.3 **Admixtures**
2.4 **Additional materials**
2.5 **Water**
2.6 **Chlorides**

Within each of the major subsections, there may be several further subdivisions, as follows:

2.10 **Ready-mixed concrete**

2.10.1 Approval of plant
Ready-mixed concrete shall be produced at an approved depot which must be certified to meet the requirements of the National Accreditation Council for Certification Bodies, Category 2, for product conformity.

All delivery tickets shall be retained by the Contractor and made available for examination throughout the duration of the Contract.

2.10.2 Additional water
All the constituents for each mix shall be added at the manufacturer's depot. No extra water or other material shall be added after the concrete has left the depot.

2.10.3 Rejected concrete
Rejected concrete shall be removed from the site. The delivery ticket shall be marked 'REJECTED'.

One or two comments may be made about this extract:

1 References to accreditation bodies are common in specifications, and act almost as additional normative references.

2 Capital letters are used to show specific defined terms—in this case, the Contractor and the Contract. The manufacturer, however, is not defined at this stage, and so the lower case 'm' is used.
3 The exact wording on the delivery ticket of rejected concrete may seem to be an example of over-pedantic detail, unless the consequences of unclear or inconsistent marking are taken into account. Specifications frequently need to be as precise as this.

So much subdivision and detail is not as confusing as it seems, since a specification is not read or even used in the same way as a report (see page 50); it is a series of descriptions/requirements which will form the basis for work, and its format—and to a certain extent its style—is less important than the ease and accuracy with which it can be used. A distinction should be made between the requirements themselves and any explanations or notes; this may be achieved by employing a different font, for instance italic for non-requirement information or bold type for general requirements which apply to detail already given. In the example of concrete given above, the following sentence might well follow a section on, for instance, Alkali-Silica reaction resistant concrete:

The Contractor shall provide evidence of compliance with this clause for proposed sources of materials and such evidence shall include a calculation of the total equivalent Na_2O from all sources within the concrete.

Sometimes a specification is produced in response to a request, and the format is laid down by the client. This is helpful to the specifier and tends to reinforce a sense of co-operation between the parties involved. Company standard formats may be available through the computer network, or earlier specifications may provide a framework for the new document. It is always worth getting second and third opinions about the format, and, if possible, the client's agreement before starting to write.

Style and usage

We wrote above that the style of a specification is less important than its accuracy and ease of use; it is, for instance, more important to repeat the same precise word than to try to find an equivalent, as one would in writing a technical article or report. A 'literary' flowing style is almost impossible to achieve in a specification: the text is too broken up, and sentences within any particular passage tend to be of the same length and structure.

There are, however, aspects of style and expression which are similar to those of report writing, and others which have especial importance in the writing of specifications. The guidelines recommending simple words and short sentences as far as possible (see page 20) are the same; avoiding jargon or colloquialisms is perhaps even more important, as many specifications are produced for an international readership whose first language may not be English.

This constraint should persuade the specifier to avoid over-complicated forms of address. Directions such as 'on no account', meaning 'do not', or 'whether it be in respect of ... or', meaning 'either ... or', are dangerous: a real-life illustration of such hazards tells of the instruction 'Under no circumstances switch this off' reaching an end user whose command of English was not good. He puzzled over the meaning for a moment or two, and then said to his colleagues, 'I'm not sure about the first few words, but the main message is clear enough—"switch this off"!'

Specifiers may also be tempted to employ pseudo-legal language, which poses the same problem for the overseas reader. Expressions such as 'aforesaid', 'hereinafter' and 'provided always that' can almost always be replaced by more normal English. Words may also be heaped up in an unnecessary way—'orders, instructions and directions', all meaning the same thing.

Words should, as far as possible, be both simple and precise. 'Regular maintenance' sounds clear, but how regular is regular? Once a week, once a year, once every ten years? 'An equivalent product' begs the question '*How* equivalent?' Under some circumstances, a pen, a pencil or a computer may be equivalent agents with which to write a book—but the authors and publishers of this book have no doubt about the differences! Terms such as 'workmanlike' or 'matching' lead to similar uncertainty, as may expressions which do not travel well: 'The operation of the system may be affected by bad weather' seems clear enough in a British context, but in some parts of the world 'bad weather' may mean drought or excessive heat. All such expressions must be avoided or defined more precisely.

The word 'must' is rarely used in specifications, but it is one of a group of words which cause great problems to the specifier. In ordinary life, we use 'will' and 'shall' more or less interchangeably, relying on the inflection of the voice to show what we mean. 'I will go ... ' has a different impact from 'I *will* go ... ', and the difference is shown in writing, as here, by the use of italics or perhaps an exclamation mark at the end of the sentence. Similarly, 'I shall go ... ' implies a simple future, while 'I *shall* go ... ' suggests that there is opposition which will be disregarded.

In a specification, 'shall' implies obligation—this *has* to be done—and must not be used in any other sense. The use of 'must' should be reserved for what is mandatory quite apart from the requirements of the specification, for instance because of British or European law or regulations (such as Eurocodes). The distinction between 'shall' and 'must' is, unfortunately, not always observed, in spite of the support of the relevant British Standard (BS 7373, Section 2: Presentation of Specifications).

In practice, the use of 'shall' causes less difficulty in a specification, where its meaning is widely accepted, than in a report or similar document, where it is not so clear. If a report writer wishes to use 'shall' to refer to what is mandatory, then it is important that this usage is set out clearly in the introduction, so that the reader is not left in any doubt about its force. 'Will' usually implies a future action, as in ordinary speech, but when it has no particular force, it is better avoided. 'When the switch is pressed down, the light will come on' is a common form of writing in instructional text; the reader is left with two areas of uncertainty: is there a delay, and if so, for how long; and what is the appropriate action if the light does not come on? It is much clearer to say, 'When the switch is pressed down, the light comes on.'

Other related words cause much less of a problem. 'Should' is used for guidance or suggestion, when there is no obligation involved; 'might' refers to a possibility; 'can' and 'could' are used to suggest the capability for action—it *could* be done, but may not be, for a variety of reasons. Expressions such as 'This is to be carried out', 'It is important that', or 'This is normally required' have no clear force, and are better avoided.

Wherever possible, the active voice is preferable to the passive, especially since it is less likely to lead to uncertainty about who is to do what. In this example:

The owner of the vehicle should check the tyre pressures and the oil level. The garage should check the brake fluid. The battery should also be checked.

the first two areas of responsibility are clear, but it is not obvious who should check the battery. Such ambiguity would be dangerous in a specification. The active voice also makes a stronger impact—it suggests action—while the passive, as its name implies, is more descriptive and usually needs more words for its expression.

The punctuation of a specification is of particular importance, as bad punctuation can create ambiguity and even a wrong meaning. Many people hesitate to use a comma with 'and', often remembering from early schooldays that in a list, 'and' takes the place of a comma between the last two items. This is of course correct, but 'and' is used both to join items or actions and to show separate groupings. This latter use is often combined with 'or', and only the use and position of the comma will make the meaning plain, as in the following example.

All circular ducting shall be mitred or jointed and welded.

This might mean either:

All circular ducting shall be mitred, or jointed and welded.

or:

All circular ducting shall be mitred or jointed, and welded.

The comma is essential. 'Either ... or' constructions must also be checked to ensure that the correct information is found at each side of the 'or'; the following real-life example shows what can go wrong.

Air entry into the unit shall be effected either by diverting supply air from the associated running air handling unit or another similarly rated running air handling unit.

The balance is clearly wrong, since 'by diverting supply air' belongs to both the 'either' and the 'or'; the sentence should read:

Air entry into the unit shall be effected by diverting supply air from either the associated, or another similarly rated, running air handling unit.

Lists can be extremely useful in showing a range of information in a clear and easily read form, provided that the intention is clear. Does the following list, for example, suggest that all of the materials must be used, or that any one of them is acceptable?

Manholes shall be constructed in the following materials:

a) brick
b) concrete
c) pre-cast concrete
d) polypropylene

Revision

Inevitably, parts of specifications go out of date and have to be revised, indeed, British Standard 7373, quoted above, recommends that all specifications are reviewed within five years. This process is obviously made more simple if the text has been bound in a loose-leaf form such as a ring binder, but amendments may have to be sent out page by page until the appropriate time for revising the whole document. This means, of course, that details of all known recipients have to be kept and updated.

A report often contains cross-references from one section to another, but this is hazardous in a specification. As far as possible, the material should be self-contained, so that a revision of one part does not result in out-of-date information in another.

Consistency, at every stage of specification writing, is essential, and there must be close co-operation between all the members of the team responsible for producing the document. This applies both to checking the specification before it is issued and to checking revisions. This important subject has been dealt with in detail elsewhere (see page 31), but it is worth reminding all specifiers that a mistake, of any sort, is potentially dangerous for users and highly damaging to the credibility of the specifiers.

Key points

- A specification must be clear, unambiguous, and easy to follow and apply
- Identify the operation to be specified and the user's level of knowledge and responsibility
- Clarify the scope of the specification and plan the contents list
- Distinguish between normative and informative references, and decide the priorities involved
- Keep a record of defined terms and other definitions which will be listed in the specification
- Follow prescribed or standard formats as far as possible
- Use simple precise words, and short sentences as far as possible
- Use terms such as 'shall' and 'must' accurately and consistently
- Avoid mistakes of punctuation, which can create ambiguity and misunderstanding
- Always check specifications with the utmost care
- Record details of recipients, as most specifications are updated and revised at regular intervals

DOCUMENTATION FOR MEETINGS

Professional people spend a great deal of their time at meetings, and meetings generate a great deal of information. The first stage, often an invitation to attend the meeting giving date, time and place, may be transmitted by e-mail or telephone, as may the response. Some meetings are more or less impromptu, perhaps to tackle an immediate work problem, and the people attending may be invited simply by a colleague calling in at the office; others may involve representatives from different companies, who have to signal their intention to be present days or weeks before the meeting is due to take place.

However formal or informal the meeting, two documents are almost always needed: an agenda, setting out what is to be discussed, and minutes, recording what happened. Either may take the form of a handwritten note or a short memo; either may be a formal document which will be kept as a permanent record of the meeting.

The agenda

Almost every meeting needs an agenda if time is not to be wasted, and it must be circulated a sufficient time before the meeting to allow people to reply; this time lapse will vary, of course, but in principle it should be long enough to enable them to keep the time free in their diaries but not so long that they forget or ignore it. People often need encouragement to attend meetings—few people do so regularly with any enthusiasm—and the agenda helps by giving them advance information about the issues to be discussed. Calling people together is always an expensive business even if they work in the same building, and they need to know why their work is being interrupted in this way. They also need to prepare for the meeting, by collecting information which might be of use and by focusing their thoughts so that they are ready to listen and to contribute as appropriate, without time being wasted when they are all together.

The agenda may simply be a message: 'Please would you come along to my office so that we can discuss ... ?' or 'We need to get together to sort out .... How about tomorrow about 10.30 in my office?' Such messages give the essential information: why, when, and where, and this may be sufficient.

For a more formal meeting, much more is needed, and drawing up the agenda may itself require a meeting of the officials, especially whoever is to be in the chair, and the minute writer. The traditional ordering of an agenda demands that apologies for absence are recorded, the minutes of the last meeting read, and matters arising from those minutes discussed, before any new business. Some meetings have other 'fixed' items, such as Chairman's or Treasurer's Reports. At the other end of the agenda will be 'Date of next meeting', which must be recorded so that everybody is aware of it, including those who missed the current meeting.

Between the beginning and the end of an agenda, the items to be discussed must be set out in a logical order. This may require considerable thought if the meeting is not to backtrack, going over early decisions again because later discussion is suddenly seen to have a bearing on earlier matters. There is debate over whether it is better to get trivial items over quickly before settling down to the more important matters, or to start with the weightier discussions while people are comparatively fresh and willing to concentrate. The latter system has the added advantage that if someone has to leave early, it is the trivial items that are missed. However, some decisions may be urgent but not worth spending much time on, and it is usually a good idea to put these early on the agenda, but to note that they will have a strict time limit; realizing that two or three decisions have already been made although the meeting has only been under way for 15 minutes gives encouragement to everyone.

Many meetings last far too long: those present become bored and feel that they are wasting their time, and as a result of this, the discussion becomes even more protracted and perhaps more irritable in tone. It is often helpful if the agenda includes not only the time to be spent on the trivial items but also the time at which the meeting is scheduled to finish. Realizing that time is running out, and that if decisions are not made soon it will be necessary to hold a further meeting, concentrates the minds of those present in a most useful way. Knowing when the meeting will end is also an incentive to attending it!

The agenda should repeat the date, time and place of the meeting, and should inform readers whom to contact to say whether they will be there or not. It is of course sometimes a courtesy to enclose a map or other directions for finding the venue.

The minutes

Some meetings have to be recorded formally, while others produce no more than a brief memo giving the decisions reached. A record of some kind is always needed, however, to ensure that agreed actions are carried out and that other people are informed that the meeting took place and that decisions were made. The minutes also show that the decisions were fair ones, at least in that they were not made unilaterally but after discussion and agreement (or at least discussion). If the meeting takes place regularly, the minutes form a basis for the following meeting, and so become a kind of historical record of company activity.

At the first meeting of a series, decisions have to be made about the minutes. How much should be recorded? How far should individuals be identified, and how are actions to be linked to the names of those who have to carry them out? It may be sufficient to record decisions, or to summarize discussion without naming individual speakers, or it may be wise to record a great deal of what is said, showing in detail how agreement was reached. These decisions will have far-reaching consequences, as precedent is the most important guide to what is recorded, and once a particular pattern of minutes has been established, it will be difficult to change it.

Action columns are extremely useful. A margin is drawn down the right-hand side of the minutes, and in it are recorded the names or initials of the person (or company) responsible for implementing the decision, and the date by which it is expected that the action will have been completed. At subsequent meetings, this column can be checked during 'Matters arising', and the action can either be ticked off as completed or carried forward to the next meeting. This system identifies actions and responsibilities clearly, and prevents their being overlooked.

Few meetings require a full record of all that is said: it would be unwieldy and is usually unnecessary. If the minutes contain a summary of the main suggestions, questions or disagreements, with a full record of decisions and actions to be implemented, recorded in the order in which they occurred, that is usually enough. Over-long minutes, naturally enough, tend not to be read.

Minutes are usually numbered, for easy reference. This may take the form of a year followed by a meeting reference, followed by the minute number, so that 96/3: 2 indicates the second minute recorded at the third meeting (in the case of a monthly meeting, in March) of 1996. This is sufficient for meetings which happen on a monthly or less frequent basis, but perhaps not ideal for those which take place more often. The other commonly used system uses headings, with subheadings if necessary, followed by paragraph numbers, for instance:

3   **New car park**
3.1   Security
1.
2.
3.

This makes cross-reference easier, as the subject under discussion is highlighted; it is clearly useful when a meeting covers several subjects. It could, however, be too complicated for an informal meeting.

The actual events of the meeting are preceded by information which is more or less standard. The meeting must be identified in some way, perhaps by both name and number; where and when it took place, and the names of all those present should be clearly given. Officers, for instance the people taking the chair and recording the minutes, should be identified, and if the meeting is one of a series, that too must be made clear. Such information may be printed in bold type at the top of the first page of the minutes.

It is usual for minutes to be circulated to those who were present and to those who were invited but were unable to attend; they may go also to other interested parties, or be displayed on a notice board for general information. At the beginning of the next meeting, they will be read (or taken as read) and signed by the person taking the chair; this gives them the force of a legal document which may be referred to or quoted in the future. It is therefore essential that minutes are written clearly and unambiguously.

Style of minutes

The requirement just given for clarity and accuracy is an overriding one. Considerations of good writing style are unimportant; minutes may be written in note form as long as the meaning is clear. They should of course be correct in spelling, punctuation and grammar, or they will give a poor impression of the writer (and may be difficult to read or ambiguous).

There is, however, one essential requirement of minutes: they should be written in a neutral way, unemotionally and objectively. They must reproduce as closely as possible the words spoken, and the minute writer's job is simply to record. It is perilously easy to add to the meaning simply by the choice of inappropriate terms. For example, 'he said' might be recorded as 'he stated', 'he emphasized', 'he declared', and what was perhaps a casual comment has become almost a matter of principle. It is better to repeat the same phrase, against the demands of good style, than to change it and imply more than was intended. It has been said that the minute writer is effectively in control of the meeting, and this is true in that posterity will know only that which has been recorded in the minutes—nothing else will remain of the meeting. Yet the minute writer is also the servant of the meeting, with the responsibility of ensuring that what passes into history is an accurate reflection of what took place.

Key points

- An agenda gives advance information about the subjects for discussion, and helps people to focus their thoughts beforehand
- The items to be discussed should be in a logical order, with a time limit if appropriate
- Knowing the time at which the meeting will end encourages people to attend
- Precedent will dictate how much of the meeting is recorded: always record the decisions
- Action columns identify actions and responsibilities clearly, so that they are not overlooked
- Standard information about the officers of the meeting is given at the start of the minutes
- The minutes are numbered for easy reference; they must be written clearly and unambiguously
- If you write the minutes, you are responsible for reflecting accurately and without prejudice what was said at the meeting and the atmosphere in which the discussions took place

PROCEDURES AND INSTRUCTIONS

All organizations have some kind of procedure, or work would never be completed. Some procedures remain informal and unwritten: the people involved know what has to be done and do it, and it is assumed that anyone new to the organization would have sufficient knowledge and experience to be able to follow what everyone else has been doing.

Nowadays, however, that is rarely sufficient. Procedures are written down and so formalized; they are audited, sometimes by outside authorities, and if they do not present a detailed and accurate picture of the work carried out, they fail the test and have to be revised. In any case, procedures need regular revision, as work patterns change—tasks are added or deleted, areas of responsibility are reassigned—and an out-of-date procedure is not only of little use but also positively misleading. A procedure is put in place as a guide for people at work, indicating the nature of the tasks to be completed and the responsibilities these entail, and those who use it must feel that it is up to date and reliable.

There are two levels of procedure: general, which give an overall view of what should be done in terms of normal working practice, and specific, which show the detail of particular work activities. Strictly speaking, neither gives orders: procedures say what should be done, but it is left to instructions, which will be discussed later in this chapter, to tell the reader what to do, step by step. In general terms, 'this is what should be done' indicates a procedure; 'do this' orders the reader to take appropriate action. However, this distinction may in practice be blurred, as some specific procedures include sections which are in fact instructions for carrying out a particular activity.

General procedures

All procedures have to take for granted a certain level of knowledge and ability. Nothing can be described in such detail that the activity could be carried out by absolutely anybody, and the first decision made by any procedure writer is about the user: who is going to be responsible for this task, who will carry it out, what do they already know and what do they need to be told? In the case of a general procedure, the answers to these questions may be wide—a range of people may be involved in carrying out the activity under various circumstances. An example from *Lloyd's Survey Handbook* (6th edition, edited by Bryan Lower-Hill and published by Lloyd's of London Press Limited, 1996) shows this in practice. In an emergency involving dangerous goods carried by sea, the priorities of a marine surveyor must include the following:

(a)   Immediate first aid treatment to any personnel affected.
(b)   Evacuation—clearing from the site all persons other than those directly involved in making the situation safe.
(c)   Ensuring that any goods, materials, containers or equipment which may have been contaminated are not removed from the site until decontaminated and declared safe.
(d)   Notification to the appropriate authorities ...

This extract is a part of a general procedure. No details are given, for example, about what sort of first aid treatment should be given or who the appropriate authorities might be. Circumstances would make these clear, and no doubt precise instructions would be given to those responsible for taking action.

The language reflects the generality of this procedure—nobody is instructed to do anything, these are *among* the relevant priorities, and the style is passive ('ensuring that ... are not removed'). These priorities are a general guide to actions which should be taken if a particular type of emergency should arise.

Such general procedures reflect the working practice of a company or organization. In preparing such a document, the user must be identified (in the case of the example above, the marine surveyor involved in assessing the problem), the level of knowledge clarified (a marine surveyor would be expected to know who the appropriate authorities are, if an emergency should arise), the purpose of the procedure defined (safety, in this case), and the areas of responsibilities allocated (these are the responsibilities of the surveyor).

Specific procedures

Beyond the general procedure, there are specific procedures for carrying out particular tasks. These describe in detail what is involved in the work, showing how, for example, a machine should be set up, operated, maintained and decommissioned. Identification of the user is much more precise, and of great importance in assessing the level of detail given. The user's knowledge and experience are critical: if too little detail is given, the operation cannot be fully carried out; if too much is given, the user may become bored, irritated or inattentive, and may then disregard the procedure altogether. It is tempting to give as many details as possible so that nothing is overlooked, but if the user is already familiar with the general principles which apply, the result can be patronizing and distracting. For this reason especially, the user's level of knowledge must be assessed accurately.

A simple procedure may be listed in one sequence, either numbered or lettered. Among the information given to marine surveyors, quoted above, is a short procedure prepared to guide a surveyor who has to check a cargo of vinyl roll floorings; it is written not for the packer but for the person who might need to investigate the mode of packing if a problem were to arise. It includes the following:

Vinyl roll floorings
(a) Rolls should be stored in the factory packaging as received.
(b) The area where the rolls are to be stored should be dry and the floor area should be level.
(c) Rolls must be free standing and in an upright position. They should not be laid flat or have any external loads bearing against them.
(d) The store temperature should be between 16 degrees C and 32 degrees C.

The needs of the user have been carefully assessed in this passage. It deals with a specific question, from the point of view of the person checking; it does not give instructions for the person who has to carry out the storage. In some ways it is a checklist of what to look out for if damage has occurred—a very basic type of procedure. The language reflects this: 'should be', 'are to be', 'must be' are used interchangeably, as there is no particular reason to stress differing degrees of urgency within the procedure.

Layout

However simple or complex the procedure, the stages must be presented in a logical sequence, either in words or in a diagram such as a flowchart. Each step must be described precisely, in simple language as far as possible, with space and headings used to clarify the connections between stages of the procedure. Useful notes (for instance, a warning about the possible dangers of the action to be taken) may be included. As in a specification, a different font, perhaps italic, might be used for comments as distinct from main text, or a different layout on the page might show the user that notes for guidance are included at that point.

The numbering system of a procedure might resemble that of a report (see page 58), with the main headings and their attendant arabic numerals highlighting the major stages, and subheadings and subordinate numbers indicating the minor actions within each major stage. Notes and comments should obviously be kept outside this pattern.

A complex procedure is likely to have title and contents pages as in a report, and to include standard sections, such as Purpose, Scope, Revision History, Related or Relevant Procedures or Documents, and Responsibilities. A section giving instructions might also, as suggested earlier, be included. The following example shows how the early sections of a procedure might be set out.

1 PURPOSE

The purpose of this procedure is to measure a differential pressure between two adjacent rooms, or a pressure drop across an air filter, using an Electronic Micromanometer, an Analogue Magnahelic Gauge or an Inclined Fluid Manometer.

This is to ensure that:

1.1 Air moves from an area of higher pressure to one of lower pressure.

1.2 The differential pressure across a filter can be monitored, therefore determining its condition.

2 SCOPE

This procedure applies to the measurement of differential pressures in manufacturing areas and associated air conditioning plants.

3 REVISION HISTORY

This procedure supersedes xxxxxxx.

4 HEALTH, SAFETY AND ENVIRONMENTAL PROTECTION

The work associated with this procedure is carried out in production areas and engineering plant rooms. A permit to work must be obtained from the production or engineering section head to ensure that staff are aware of locally enforced safety procedures.

5 RESPONSIBILITY

5.1 The monitoring of differential pressures and pressure drops across a filter should be performed only by company staff or company-appointed contract staff trained in this procedure.

5.2 All persons carrying out this procedure must ensure that the equipment used has a current traceable calibration certificate.

6 REFERENCES

There are no references for this procedure.

(These introductory sections are followed in the procedure by detailed instructions.)

The last section quoted might at first seem a little odd, but it is useful in a procedure to state that there is no relevant information, as this reassures the user that there has been no oversight.

The layout of this example is helpful. The headings stand out clearly, and there is a good space between sections. The numbering system is commonly used and logical, and the subsections are indented. Where instructions are also given, they can usefully follow a similar pattern.

Instructions

Instructions are issued either within a procedure or as a separate document. In a sense, they complement procedures, by giving the user stage-by-stage directions so that the activity described in the procedure can be carried out.

If instructions are issued as an independent document, they can have a similar structure to that of a procedure, as Purpose, Scope, Responsibility, etc. may in some cases be appropriate sections. However, safety is of paramount importance, and while general warnings should be given at the beginning of the document, specific warnings must be given at the start of the particular step to which they apply. These should be highlighted in an appropriate way, by the use of bold or italic type, extra space, boxing or any similar device which will draw them to the user's attention. Such devices must be used consistently, as should expressions such as 'care needed', 'warning', 'hazard' and so on: the precise meaning must be defined, as there is no generally accepted hierarchy which would ensure that 'danger' is more or less extreme than 'hazard'.

Style of instructions

The most important difference between a procedure and instructions is in the language that is used. Words such as *will, shall, may, could* have no place in instructions; the imperative should always be used, for instance 'Switch off the engine' rather than 'The engine should be switched off'. Vague terms such as 'regular' or 'large' should be avoided, as should other imprecise terms such as 'front', 'back', 'near', 'top' or 'bottom'. The style of instructions should always be as simple and direct as possible, with diagrams to clarify the detail whenever possible.

Organization

Instructions must always be given one step at a time, each being clearly numbered. So, in the example of a procedure given above, under a heading 'Differential pressures between rooms', the user finds:

4 Place the high pressure tubing into the room suspected of having the highest pressure. Ensure that the door is closed but not trapping the tubing.

5 Retain the low pressure tubing in the adjacent reference room.

6 Take the reading.

7 Repeat to verify the result.

8 Record the results on a differential pressure survey diagram of the manufacturing centre.

Testing and revision

The ultimate test of both procedures and instructions is in their use, and they should always be tried out on colleagues. It is important to have instructions, especially, tried out by someone who knows enough to be able to carry out the activity but who has never before had to do it. Such a person is less likely to make assumptions, work on experience rather than instructions, or miss out stages that seem unimportant, than someone who knows only too well how the task should be performed. If this test can be carried out by one of the people who will have to use the instructions in future, there is the additional benefit for the writer of being able to check that the level of information is correct.

Procedures and instructions should be regularly revised, and their revision history made clear on the text. If a document is perceived as being out of date, it will be ignored. If it appears to be clearly written and helpfully organized, it is likely to be accepted and put into use. As with all technical writing, the goodwill and co-operation of the reader or user is essential for the success of the operation.

Key points

- A procedure is a guide to work activity and responsibility; it must be seen to be reliable and up to date
- Identify the level of knowledge and experience of the users of the procedure
- A general procedure gives guidance to the organization's working practice
- A specific procedure must be written precisely for those who will use it; it may include instructions that they must carry out
- The stages of a procedure must be set out logically, in an appropriately numbered format, with notes and comments clearly separate from the main activities
- Instructions must draw the reader's attention to information about the health and safety implications of the activity
- Instructions tell the user what to do; they should always be written in the imperative
- Try out procedures and instructions, preferably on someone who will have to use them
- Procedures and instructions must be updated at regular intervals

5
Writing for Publication

First thoughts • Publication in journals • Conference papers • Writing reviews • Writing books • Avoiding prejudice • Literary agents

Nowadays, there is a tremendous outpouring of technical publications in the form of books, journals, proceedings of conferences, material from professional institutions, and specialist articles in the quality press. All clamour for our attention—in competition with information which comes online, through e-mail and so on. Such material all has to be written, and although a good deal is produced by professional journalists, much is also the work of technical specialists, writing either as an optional extra to their everyday work or, in the case of academics, as an important part of what they are contracted to do.

This chapter is intended for such specialists, and especially for those who have never written for publication before. It is a daunting prospect: most publications look authoritative and often impressive, and many are extremely expensive; new writers sometimes feel that there is a great gulf between themselves and established authors.

Yet such would-be writers are themselves part of the market for technical publications, and they have probably assessed a range of books and articles for their own convenience, deciding which present new ideas, which are useful textbooks, which are especially up to date, and in which they would like to see their own work appear. This is extremely useful background research, as writing without a focus is rarely successful—the work most likely to be published is aimed at a particular market and even at a specific journal or publisher.

FIRST THOUGHTS

First of all, the writer needs to analyse what he or she wants to say. Are the ideas original? Do they constitute a response to earlier published work? Is the material suitable for a major refereed journal? If so, there will be constraints: because of the reviewing which will take place, an article is unlikely to be published quickly; the editors may receive far more material than they could possibly handle even if it were all worth publishing, and much of the journal itself might well be written by a small number of established specialist writers. Nevertheless, having an article accepted in such circumstances brings status, perhaps promotion, maybe international attention.

There are also more 'popular' technical journals. These are often published weekly or monthly, and so the wait for publication is likely to be shorter; since they use so much material, they are more likely to look kindly on new writers whose material is presented appropriately. However, they are not generally interested in the results of research unless it has a clearly-defined practical outcome, and—perhaps sadly—publishing in such journals is not considered prestigious in academic circles (although information may thereby reach many influential people in the industry who would never read an academic journal).

Each writer has to consider which is the appropriate type of journal for the particular information or idea; academics especially may publish from time to time in both refereed and more popular journals. However, it is generally true to say that the more the individual publishes, the easier it is to get material accepted, and the first publication may well present the biggest challenge. In view of this, a sensible place to start may be in the letters section which most journals carry: a comment on a previous letter, in agreement or disagreement, perhaps a personal experience or a question addressed to an earlier contributor—all these give the writer experience in producing material clearly and concisely, and appropriately for the journal's readership. Such contributions should be of the length usually published (or shorter) and to the point—editors will be quickly put off by material that rambles or seems to address too many points at a time. If the letter is published and further correspondence develops as a result, so much the better; the editor will remember, and the writer will have been noticed by other specialists in the field.

PUBLICATION IN JOURNALS

Once the decision about the appropriate type of journal has been taken, the next step is to use the university or company library to find two or three journals which seem to cover the same field. The writer's own experience, and that of colleagues, will be of assistance: where have useful articles appeared in the past? Which are the most prestigious journals in the field? When the writer feels that a particular journal stands out from the others, it is important to read at least two back numbers, as a special issue might give a misleading impression of the usual coverage.

The writer will be looking for the answers to several questions:

1 How long are the articles? (Count the words!) Does there seem to be a pattern of a few very long articles followed by a wider range of short, 'taster' features?
2 What is the style of the writing? Does it seek to attract a wide range of intelligent interest, or is it very formal and obviously written for a small number of highly specialized readers? It is essential to identify the target readership. (There may be a short paragraph in the journal itself which clarifies this point—it is worth checking the page which gives the names of the editorial staff.)
3 What is the layout? Are the articles printed across the page or in columns? This will affect the writing style: sentences and paragraphs should be shorter if they will appear in columns.
4 What types of diagram are used? Are they printed in colour, or in black and white? How are they labelled and/or numbered?
5 What is the format of references (if any)?

This information will help the writer to plan an article which fits the style of the journal—and which therefore shows awareness of what is needed. The fewer changes the editor has to make, the more likely the article is to be accepted.

Such advice seems obvious, and yet many writers produce the article they want to write without reference to the editor—and then are surprised when it is rejected. An article which is 'all-purpose', written by the author with no reference to any specific publication, is not likely to impress an editor. Making contact first is always worth while. A short letter giving a brief outline of the proposed article, with a note of its relevance, and of its suggested length, will be helpful to the editor in making a decision, although if the writer is completely unknown, there may be only a provisional go-ahead. The editor may reserve final judgement until the full text is available. An initial letter proposing the article can be followed up by a telephone call, but not too quickly or too often—it doesn't pay to push the editor too hard!

The article should be written in an appropriate style. If the journal is aimed at industry rather than academics, the main message should appear in the first paragraph and be supported by good photographs. A busy manager will probably glance at the page and pass on, unless a clear message attracts the attention. This is very different from academic readers who will give time and effort to reading material in their chosen fields. In both cases, however, always write just short of the agreed length. It is easy to fill space, with an out-take or with boxed material, but if someone else has to cut the text down to size, the author's prized last sentence may completely disappear!

As with any important document, the article should be shown to at least one knowledgeable colleague for comment before it is submitted. A long wait is likely to follow, as the article is assessed by the editor or editorial panel in the light not only of its own qualities but also of how much material is available for publication at the time. There may eventually be one of three responses: the article may be accepted as it stands, which is the best possible news; it may be accepted subject to amendment; it may be rejected, with or without helpful comment.

If it is accepted, page proofs may follow, although if it is to be published quickly the author may not have the chance to check the text again. In the case of a refereed journal, proofs are almost certain to be sent, with notes about the marking-up and a date for return—which should be adhered to if publication is not to be delayed further. After this stage, the author must wait for the finished product; editors will not take kindly to major changes at proof stage and even less kindly to any changes at all after the proofs have been returned.

Reviewers who are specialists in the field may have made suggestions which the editor would like to see incorporated into the text. The new author has to remember that these are experts, and that the changes may make the article more acceptable as a final product; at the same time, the author has some rights, and if necessary can object to the changes, with courtesy and reasoned argument. Needless to say, the editor has the last word unless the author wants to withdraw the article completely!

Rejections are common and hard to bear from the writer's point of view. If the editor has taken the trouble to make helpful comments, this is encouraging: editors are busy people and it is much easier simply to reject. The implied message is probably that the article has potential; maybe it is not quite right for the particular journal or perhaps the timing is not appropriate, or it may be that something similar has recently appeared elsewhere—for all kinds of valid reasons, articles are turned down. The best advice for the would-be writer is: take all the advice you can get, and try elsewhere (or write a different article). Giving up in despair is not a recommended course of action.

Finally in this section, we can take as an example of important refereed journals the *Proceedings* of the Institution of Electrical Engineers. There is a range of *Proceedings* for specialized interests, such as *Vision, Image and Signal Processing, Electric Power Applications, Control Theory and Applications* and *Radar, Sonar and Navigation*. From time to time, there are also special issues devoted to particular topics of interest: these are issued as stand-alone publications. The IEE *Proceedings* are produced in printed and microfiche form, and are published bi-monthly, each part being indexed by subject and author.

This is basic information which we may assume any specialist in the fields covered is likely to know. The hopeful writer will also need to know that all items which appear are peer-reviewed and selected by expert honorary editors. These are prestigious journals in which to be published, and our writer will need to know exactly what is involved: the IEE will send further details about each, and the names of the honorary editors, together with an authors' guide sheet.

As an example of the type of guidance given to authors, we quote below, with permission, the first section of the sheet provided by *Electronics · Letters*, one of the IEE's refereed journals.

GUIDE TO AUTHORS

1   *Electronics Letters* is intended to provide a rapid means of communicating new information and results on important topics of current research interest. Letters submitted for publication should therefore record original work not previously published in the open literature. Letters that are under consideration by, or have already been accepted for publication by, another journal are not acceptable for *Electronics Letters*. Letters should not be submitted that depend in any way upon work still under consideration by this or any other journal.

2   The subject field of *Electronics Letters* embraces electronic science and engineering, telecommunications and optoelectronics. Contributions of a theoretical nature should show a specific application in one of these fields. Where such contributions are mathematical, the author is expected to explain in detail the application of the mathematics.

3   A Letter accepted for publication by the referees is usually published within six weeks. The date of receipt of the manuscript is published. To maintain the speed of publication, proofs are not sent to authors, and corrections cannot normally be made once a contribution has been accepted. Extreme care should therefore be taken with the preparation of all material submitted.

4   Authors are not encouraged to resubmit declined papers. Referees are asked to bear this policy in mind when making their reports. However, minor errors will not preclude the publication of otherwise acceptable material.

5   Unclear English may influence the referees against otherwise good papers. Authors who are not native speakers of English should consult colleagues who are, or language experts.

6   Short comments on published Letters are considered for publication. The original authors are invited to reply. Neither comment nor reply should exceed 500 words nor contain more than one figure. The comment and reply will be sent to referees, where possible to those who reviewed the original Letter.

Details of the Manuscript Requirements follow these guidelines, including advice about the format and presentation of the manuscript and accompanying diagrams, units to be used and the form of references. One further sentence is worth quoting:

Please note that manuscripts which fail to comply with these requirements are liable (and manuscripts which are too long are certain) to be returned to the authors without being sent to referees.

The importance of adhering to the publisher's requirements is obvious; they are also helpful for the writer in that some decisions have already been made, and the constraints of the article are there in black and white before the process of writing begins.

CONFERENCE PAPERS

Two particular forms of article need a brief mention: conference papers which are subsequently published, and review articles. Most professional organizations which hold conferences issue a printed version of the papers given, as a lasting record of what was said and in order to reach a wider audience.

This presents a problem for the speaker/writer: the information may be written in language appropriate for publication or prepared for the conference itself, that is, the language is intended to be spoken rather than read. Ideally, notes should be established for the presentation and a separate text should then be prepared for publication. In practice, this does not always happen. It is most important that both writer and editor should be aware of the difference between the written and the spoken language, and also of the details which cannot sensibly be given in spoken form (such as references or equations). If a disk is handed in at the conference, the writer should also have the opportunity to revise the text in the light of the discussion; this gives a further occasion on which the text can be checked for accuracy of information and readability.

The habit of 'reading a paper', literally, at a conference is not recommended (see the section on conference speaking, page 131); the distinction between giving all the technical/scientific detail in a paper and letting the audience know the highlights/latest developments/benefits should always be kept in mind.

WRITING REVIEWS

Preparing a review or a longer review article is often a good way for the new researcher to get into print; the difficulty is that editors like to have a known 'name' at the end of reviews to add prestige to what has been said and indeed to the journal itself. It is, however, perfectly reasonable to make an offer to review highly specialized material—the editor may not have a suitable expert in mind. If the offer is taken up, then the questions asked in preparing an article apply here, especially the constraints about length.

It is not always realized that publishers also need reviewers, to comment on the information and the style of books being considered for publication (see the comments about writing books, below). There is no reason why someone with specialized knowledge should not contact a suitable publisher and, giving all relevant qualifications and experience, offer to be a reviewer on books which cover the area of interest. Such work is unlikely to occur often, but it is a way of keeping up to date with the latest publications in a given field.

WRITING BOOKS

A great deal of technical information is published in articles, conference proceedings and similar rapid means of dissemination. Books present a greater difficulty, as they take very much longer to write, and the time lapse between the author finishing the writing and the book appearing in the bookshop is a lengthy one—normally 12 to 18 months, and sometimes longer. Writing books is therefore not a sensible way to disseminate for the first time the results of research or new developments. Books also go out of date remarkably quickly, especially in the field of technology. They may add to the writer's prestige, but are unlikely in the long term to make a great difference to anyone's bank balance.

Yet we have all been educated and trained largely through the medium of books, and in spite of all the modern technology available, today's students still rely heavily on textbooks, and today's academic researcher will not get promotion without producing at least one book and probably more than one.

The first stage of preparing to write a book is very similar to that of writing an article: the writer has to have some sense of the appropriate readership, and then needs to look at the publishers which cover the field. It would be pointless for a large number of publishers to work in the same highly specialized area, and so they also tend to specialize: a look round the technical or scientific areas of a research library will show at once who concentrates on what. There are also differences of level: some publishers will tend to issue school textbooks, some will focus on undergraduate texts, and so on. Large-scale publishers will have different editors in charge of the different levels.

Publishers are usually interested in meeting potential writers in the fields in which they publish. It is well worth while contacting the appropriate editor and arranging a meeting—provided that the writer has prepared thoroughly for this encounter.

It is not a good idea to present the editor with a book ready to be published, indeed, it would be highly counter-productive to do so. A small number of well-known and highly-regarded authors might get away with such a move, but most writers need the advice and support of their publishers, and of the reviewers who will look at the manuscript at various stages and make comments.

For a start, the publisher will need a written proposal. This is a good point at which to begin planning, as the contents of this proposal will involve the writer in assessing the need for the book, its target audience and its selling points. If a draft of the proposal is taken to the first meeting with the editor, there is solid ground for discussion—the writer's homework has clearly been done.

As an example of the material which the publisher will need in the proposal, we give below, with permission, an extract from the guide, *Publishing with McGraw-Hill*, which the publishers of this book send to prospective writers.

Your book proposal

To help us to assess your proposal or manuscript, we need the following information.

The market

- Who is the intended reader?
- Will your book service an international market?
- A brief comment as to the various groups of potential readers (e.g. student, practising manager, adviser, computer expert) with specific details, where applicable, about the courses or required level of professional expertise.
- How large do you estimate the potential readership to be?

The content

- A synopsis of the proposed book, including a detailed outline of the work with intended chapter headings together with a description of each chapter under the headings, typically two or three pages long.
- Sample material, two or three chapters if possible.
- A rationale describing why the book is needed and any new ideas or developments you intend to cover, or new approaches that you intend to use. Notes on additional features such as case studies, checklists, diagrams, photographs, supplements, software, etc.
- An estimate of the length of the book.

The more we are provided with, the easier you make it for us to make a fair assessment of your proposal for publication.

The competition

A list of any or all competing books with price, publisher, year of publication and any other useful information, together with a comment as to how your book differs, what makes it superior and how it will compete.

Yourself

Details of yourself, your experience, and any of your previous publications (whether articles, reports or books).

It may be difficult for a new writer to assess some of these ideas, for instance the size of a particular market, and the publisher's editor will probably be able to help. It is a useful checklist, however, in focusing the writer's mind on the information which is likely to be needed at a very early stage of the book preparation.

This material will be sent to experts in the field and potential purchasers, and their input will be very helpful. They will, of course, suggest whether the book should be published or not, but they may also provide feedback on the treatment of the subject, the style and so on, which will guide the writer in the most appropriate direction.

An author's guide provides other useful information, for instance about the vexed question of copyright. The author is responsible for all permissions to reproduce material, and the publisher is likely to want such agreement in writing. Generally, the author will need permission when the number of words quoted from another source amounts to over 150 words, either in one block or added up throughout the manuscript. Permission is also needed if any diagrammatic or photographic material which is not originated by the author is to be included or adapted. This guideline may change in detail as the law of copyright is amended or updated, and it is important to discuss the details with the publisher at an early stage of the preparation of the book.

In most cases, the completed manuscript will be reviewed again, although it is unlikely that major changes will be suggested if earlier advice has been followed. Book production is a long process but a rewarding one, and it is in the interests of both the author and the publisher that the final product sells as widely as possible.

AVOIDING PREJUDICE

There are various constraints on writing which the author needs to bear in mind. Obviously, no reputable publisher would wish to be associated with seditious, inflammatory or illegal material, but there are other aspects of writing which can cause problems. Most publishers want to avoid sexism, and authors are advised not to use 'he' throughout a text, or in a way which could cause offence ('the manager ... he ... the secretary ... she' and so on). It is much easier to be aware of this from the start of the writing process than to try to change sentences round afterwards. For instance, we could have said 'the author is advised not to use "he" throughout a text, as if he did, he would cause offence.' By using the plural 'authors' and by changing the form of the sentence, the potential problem has been avoided. With a little practice, it is surprisingly easy to avoid either the use of 'he' or the cumbersome 'he or she'—though perhaps not always, as the careful reader of this book will have noticed!

There is another area of unintentional prejudice which must be avoided if the article or book is intended for an international audience. It is easy to write exclusively for one's own culture. British authors will tend to use English money, legal codes and case studies, and these can be confusing or off-putting for overseas readers. It is helpful to give equivalents in dollars or deutschmarks, to use an example or a case study in an overseas setting, or to avoid what is country-specific. British jargon can be difficult for readers whose first language is not English, while a greater use of diagrams will encourage readers who find the language of the text difficult.

Obviously much British material reaches a world market without being translated, but when the information is clearly designed to reach readers of many nationalities, it makes sense to help them as much as possible.

LITERARY AGENTS

Contrary to popular opinion, it is not only best-selling novelists who use the services of literary agents! They can play an important role in helping the author to choose the right publisher, encouraging the writing process and drawing up contracts which include all kinds of circumstances and rights that most authors would never think of.

It is probably fair to say that as agents want to make a reasonable living, they like to take on writers who have proved themselves already, by having articles or even books published. They are also not fond of the 'one book' author, who will never want to publish again. Academic writers tend to learn from one another about the most appropriate publishers, and they often earn very little in the way of royalties anyway.

This still leaves a good number of technical writers who would like to write fairly regularly, and who might produce a best-selling textbook or the standard reference book on the subject. Such writers should at least consider using an agent, for the very real support and encouragement which isolated authors often need, for advice on subjects such as getting permission for quotations, and for their skill in drawing up contracts which give the best terms available to the author.

The best way to find an agent is by recommendation, but there is always the *Writers' and Artists' Yearbook* which gives advice and lists agents (who may have specialized interests). We would add that it is important to see one's agent as a friend: there are countless times when a piece of advice or support is worth far more than the percentage of royalties which agents charge.

Writing for publication is hard work, but it can be most rewarding. A large number of technical articles and books in English are produced every year, and the pleasure of seeing oneself in print never quite disappears. Technical writers have an important role to play in the dissemination of knowledge and it is worth taking any opportunity to combine personal achievement with the very real value of the printed word.

Key points

- Assess what you want to say in the light of the journals which cover your field of interest
- Check the format and style of typical articles in these journals
- Get editorial approval in principle for what you want to write, before you start
- If possible, accept the advice of specialist reviewers
- Follow the publisher's guide to authors whenever possible and discuss any discrepancy
- Remember that conference papers are designed in the first instance to be spoken, and need to be adapted for publication
- Reviewing for journals is a good way of attracting attention; reviewing for publishers can help to keep you up to date with what is available in your field
- Books are a slow medium for disseminating new ideas, but can reach a world-wide readership
- Draft a proposal and seek guidance first from a publisher's editor and then from specialist reviewers
- Try to write for an international audience, and avoid any prejudice which might alienate potential readers
- If you intend to write frequently, consider acquiring an agent to handle the negotiations for you

6
Speaking Techniques

Using the voice • Non-verbal communication • Nerves and confidence •
Notes • Group presentations • Preparing the venue • Visual aids

An ancient Greek scholar is said to have remarked, with great wisdom, that
the reason we have two ears and one mouth is to enable us to listen the more
and speak the less. It may seem perverse to start a chapter on speaking with
a comment on listening, but they are, of course, two sides of the same coin.
Listening is indeed an important and difficult occupation—our attention
span is notoriously short—and the task of the speaker is to make listening
easy and, if possible, pleasurable. There are techniques which help, and this
chapter is concerned with encouraging good listening by improving the skills
of the speaker; the following chapter will put these techniques into context,
and consider the particular needs of a range of 'speaking' occasions.

One warning is necessary. A book can be very helpful in suggesting ideas
and techniques to try out, and in identifying bad practice which should be
avoided. It is no substitute for practical experience. Only giving a
presentation will tell a speaker what it is like to give a presentation, and
only regular practice will give that speaker the sense of confidence and
authority which is the hallmark of the professional.

USING THE VOICE

Using the voice effectively is probably the most important key to holding the
audience's attention. Most of us have had the unpleasant experience of
listening to a talk on a subject which interested us and being disappointed,
not by the material itself but by the way in which the speaker put it over.
Unfortunately, the most fascinating information in the world can become
boring if the speaker's voice sounds bored, or is inaudible or monotonous
or—though this is rare—too loud.

Some people are born with the advantage of a pleasant, attractively modulated voice; others are handicapped by a high-pitched or harsh tone which is difficult to listen to. It is possible to improve the voice one was born with, for instance by speech therapy or drama training, but these are extreme measures which most of us would not want to undergo, and which are in any case unnecessary for the purpose of ordinary presentations. What we can do is to use our voices as effectively as possible, and in this section various techniques are suggested to help speakers to project their voices in a way which will help the audience to listen with pleasure.

Audibility

A speaker's voice has to be heard; if it is inaudible, then all the other good things about the presentation count for nothing. It is essential to check that someone sitting right at the back of the room can hear every word, and colleagues and friends should be persuaded to come to a rehearsal in the venue for the presentation, so that there is no doubt about the volume required. Some voices seem to be squeezed out of a small tube, and travel only a short distance; the voice should be thrown, like a ball, towards the audience so that it travels a long way. It is useful to imagine blowing thistledown at someone—the substance itself is light and fragile, but it can be made to cover a distance if the sender takes a deep breath and concentrates on giving it a good send-off.

Breath control is essential to projecting the voice, and the breath itself should come from down at the diaphragm, be propelled by the stomach muscles and pass easily through the throat. If pressure is put on the throat itself, the result is a shout—very unpleasant to listen to, and almost certainly resulting in a rough, hoarse throat. The section on non-verbal communication (see page 99) deals with the need to relax the shoulders in order to control the nerves; this relaxation is equally important so that the speaker can breathe deeply—shallow breathing makes the speaker gasp for air and dries the mouth. If more volume is needed, pressure should be put not on the delicate and precious vocal cords but on the muscles of the diaphragm; as with a tube of toothpaste, if you press at the end more paste comes out at the top!

If there is a real problem with creating enough sound, the speaker must spend regular time in practising, at first in a small room and later in a bigger space, using words which are themselves explosive in sound (bang, push) so that the voice is naturally thrown out towards the listener. It can also be helpful to sing a short phrase, as some people can make a bigger sound in singing than in speaking. Confidence is essential in using the voice as it is in every other aspect of making a presentation, and anyone who has difficulty in projecting the voice needs to start in a small way, speaking to a little group of friends, and gradually pushing out more sound in a bigger room. There are a few people who will always find it difficult to project their voices, but most of us can with practice and growing confidence make ourselves heard most of the time.

Correct breathing, incidentally, will help to prevent a problem faced by anyone whose natural speaking voice tends to be high-pitched (it is a difficulty faced especially by women speakers, for obvious reasons). Nerves make the human voice rise in pitch, sometimes almost to the point of a squeaky sound, while it is generally true that a low-pitched voice is more pleasant to listen to and sounds more authoritative. Relaxing the shoulders and breathing deeply will largely overcome the problem, allowing the voice to settle at the pitch natural to it.

There comes a point, of course, at which no speaker can fill the required space with sound, and a microphone is needed. Sadly, far too many speakers resort to such aids when they could manage without help if they projected their voices effectively. Microphones are not easy to use, and many systems distort the voice and create a barrier between speaker and listener. A static microphone tends to keep the speaker rooted to the spot, which looks and feels uncomfortable; a lapel microphone is much better, although the flex can be a nuisance. As usual, practice is important so that the speaker can feel at ease with the microphone and can to a certain extent forget it. It is, of course, important to use a natural conversational volume—if the voice is projected into the microphone as on a stage, the audience will be blasted out of its seat!

Pace

Audibility is not just a question of volume; it is also the result of careful control of pace. Nerves make us speak more quickly, and in the setting of a presentation this can be exaggerated until the audience is left gasping, totally unable to keep up with the flow of words. The speed at which the presentation is delivered should be determined by the size and style of the surroundings, the relationship of speaker to audience, and the content which has to be put over. Shortage of time is never a factor.

Obviously, the speed used in a small room to a few people will be very different from that appropriate to a large hall and a big audience. In almost any presentation, the appropriate pace is likely to be slower than that of ordinary conversation: the speaker is in a sense performing rather than chatting, and every word must be clear. The rule is: the bigger the room and the larger the audience, the slower the pace. This may feel highly unnatural to the speaker but it ensures that the audience can listen with ease.

Both people and furnishings absorb sound: an empty room is much easier to speak in than a crowded room, and allowance must be made for soft furnishings such as curtains and thick carpets. It is obvious that speaking to an audience has to be a much slower process than taking part in a conversation with a friend, and the material has to be assessed accordingly. The importance of selecting the amount that needs to be said, as opposed to written, is dealt with elsewhere (see page 128); one of the main reasons for this selection is that so much less can be said effectively in a limited time than most inexperienced speakers would ever guess.

An audience needs time to adjust to a new speaker (see page 128), and the speed of the introduction is therefore particularly important, especially if the speaker has a strong accent or is speaking a language which is not primarily that of the audience. In such a case, the speed must be held back even more than usual, and the speaker should make extra use of visual aids and handout material (see page 116). Other considerations must be taken into account—has the audience listened to this speaker before and so had the opportunity to get used to the voice? Is the audience likely to be familiar with the content and to agree with it? Has the audience come simply for entertainment or casual interest, so that if a few words are missed, no lasting damage will be done?

This last will never be true, of course, of a technical presentation, when the complexity of the material will require close attention on the part of the audience and so an even more careful control of speed on the part of the speaker. Nevertheless, any presentation has shades of complexity within it, so that the speaker will at one point be describing a highly technical process, perhaps with a visual aid, and at another point perhaps be making a throw-away comment about some aspect of his or her own experience or even adding a touch of humour for a moment's light relief (but see page 129 for the appropriate use of humour). Variety of speed is helpful, so that when the speaker slows right down to describe the process, the audience instinctively feels that this aspect needs extra concentration, while the increased speed of the throw-away comment suggests immediately that this is of less immediate importance.

All this variation of speed can be planned. It is useful to record the talk onto a cassette recorder and then to listen to it, noting the points at which more speed might be possible, or when it would be helpful for the speaker to slow down or even to repeat a few words.

Highlighting

Some expressions are by their nature difficult to assimilate simply by listening. People's names are an obvious example—most of us go to social gatherings at which we are introduced to newcomers but how many of us can remember their names five minutes later? If it is appropriate and feasible, for instance at a small meeting, it is sensible to use name plates (lapel badges are usually an embarrassment, as people peer at one another or try to read the name without being seen to do so). At a larger gathering, the audience may well have a programme with the speakers' names and the titles of their contributions. This is a helpful start, but no more. If the audience has never seen the speaker before, how do they know that this is the right person? There might have been a programme change, the expected speaker might be ill, they might be in the wrong room ... they want confirmation, and as they might wish to make a note of the name, it should be spoken clearly by the chairman or the speaker; if the presentation is made by a group, then all the names must be given, and it should be made clear who is who.

Technical or scientific names may present similar problems, as may abbreviations, and in such cases the speaker has to assess whether the audience will be familiar with the term or not—it would obviously be embarrassing to a knowledgeable audience to be treated as if they were unfamiliar with a vocabulary which any professional in the field might be expected to know. Brand names, however, may be less familiar, and if the speaker is introducing a new product to the audience, the name will be repeated several times and no doubt also put in writing.

Numbers—prices, measurements, quantities and the like—are always difficult to hear: the difference in pronunciation between '50' and '15', for instance, is very slight, and it is wise to repeat the number with emphasis, or, of course, to put it in writing on a visual aid.

All such troublesome expressions benefit from being highlighted—the spoken equivalent of quotation marks. The speaker reaches the number, pauses very briefly, emphasizes the word or words, and then pauses again very briefly before continuing. The spacing combines with the slight increase in volume to warn the listener that an important piece of information is coming and to give it emphasis when it arrives.

Volume, to a certain extent, goes with pace. As a most important statement calls for extra space, so it can at the same time be given a little extra volume: this will increase the emphasis, and allow the speaker, having drawn attention to the point, to return to a normal speed and volume as soon as the point is completed. These changes are slight, but they give the audience a sense of relative importance; after all, not everything said in a 30-minute presentation can be of absolutely equal importance to everything else!

The following sentence will serve as an example of such emphasis. The speaker wants to say:

> So, as you can see, our new Alpha system costs only two-thirds of the price of similar products; the maintenance is again comparable to anything else on the market, and in addition to these benefits, your staff will be able to operate it after only two or three hours' training.

The key words in this statement are *Alpha* (the name of the system), *two-thirds* (major benefit), *maintenance, comparable* (benefit rather than disadvantage), *your* (slight compliment—your staff are particularly able), *only two or three hours* (training will not put the cost up unreasonably, you can soon have this excellent system in operation).

These words and phrases will receive slight emphasis as described above (but there are, of course, variations even within the emphasis: clearly *two-thirds of the price* is more important than *comparable*). Other expressions are relatively unimportant — *as you can see*—and these can be spoken slightly more quickly, with less emphasis. This is especially true of *anything else on the market*, which the speaker will not wish to leave fixed in the minds of the audience!

The passage above will be spoken with the following emphasis:

So, as you can *see*, our *new Alpha system* costs *only two-thirds* of the price of similar products; the *maintenance* is again *comparable* to anything else on the market, and in addition to *these benefits, your* staff will be able to operate it after *only two or three hours' training.*

The ends of words and sentences can be traps for the unwary. Many words depend for their meaning on a final consonant (*rain, afraid, bite*, and so on), and it is easy to swallow this sound (thus saying *ray, affray, by*); the context may make the word clear to the listener, but the speaker cannot rely on this. The whole of each word must be pronounced, even if, to the speaker, this seems an artificial way of talking. Some local accents increase the problem: the most notorious example is probably a London/south east England glottal stop, which swallows almost every 't' sound, producing, for example, 'wha' a lo' of li'le bo'les' ('what a lot of little bottles').

The English language encourages us to allow our voices to drop at the ends of sentences (unless the sentence is a question). This is often made worse by speakers who look down at their notes at precisely the moment at which their voices naturally fall away, so that the last few words of the sentence are inaudible. The voice must be 'supported' right to the end of the last word of the sentence, so that no information is distorted or lost. Obviously, such support is given by maintaining sufficient volume, not by raising the voice and thereby turning statements into questions.

Enthusiasm and the voice

It may seem odd to conclude this section by reminding speakers to be enthusiastic about what they are saying, but enthusiasm brightens the voice and makes it more interesting to listen to. We are not suggesting that anyone should wear a fixed grin while giving a technical presentation, but the speaker who has somewhere at the back of the mind a sense of excitement at the idea of conveying this splendid, original, fascinating information to these friendly, appreciative people, will convey something of the same feeling to the audience. Enthusiasm shows itself in the voice, and enthusiasm, like boredom, is infectious.

We should like to think that professional people who make presentations are always full of enthusiasm for the information they have to give, but experience has shown that this is not always the case. The best advice is, if you aren't enthusiastic, act as if you were. Oddly enough, such acting almost always generates the missing enthusiasm; five minutes after the presentation has started, the speaker may well be surprised to realize that he or she—as well as the audience—knows that the occasion is exciting and the material brilliant.

Key points

- Check that as you speak you can be heard clearly right to the back of the room
- Good breathing and relaxed posture are essential if you are to control the volume and pitch of your talk
- Try to avoid using a microphone, but if it is necessary, speak naturally into it
- The pace at which you speak should be determined by your surroundings, your relationship with the audience and the content of your presentation
- Allow yourself plenty of time—never try to say too much or talk too quickly for the audience to follow
- Plan variations in speed and volume to give emphasis to your main points
- Watch out for difficulties—names and numbers need to be highlighted
- Sound the ends of your words and do not allow your voice to drop too much at the end of sentences
- Be enthusiastic—it will show in your voice. If you are not enthusiastic, act as if you were!

NON-VERBAL COMMUNICATION

The subject of non-verbal communication—body language—is a fascinating one. We all use such communication most of the time, although for the most part we are scarcely aware of it. We signal our agreement, distaste, boredom or enjoyment by the way in which we sit, move and change our facial expressions; if we know someone well, we can 'read' their body language and understand their frame of mind before they have spoken a word. Non-verbal communication is extremely powerful: it can override our words and undermine the impression we are trying to give, or it can reinforce our message and persuade the listener that we are genuinely committed to what we are saying. In this section, we will consider some of the most important aspects of body language as they affect the relationship between speaker and audience, but first, one crucial reservation is needed.

Body language is deeply rooted in culture, and what is acceptable in one culture may be offensive in another. The speaker who is going to address an audience of a different cultural background *must* take time to find out what is considered to be the norm and what should be avoided at all costs. The audience may well appreciate the courtesy, or at least will listen without being irritated or upset. The way the speaker sits or stands or gestures or relates to the audience may cause offence, and that will overrule the technical content of what is said.

In this book, we will assume a western European, specifically British, audience; our advice in this section should not be taken out of that context.

Eye contact

The single most important aspect of non-verbal communication is to look at the audience (to reinforce what we have said above, this can be offensive in some other cultures, especially for a woman presenter). Eye contact is essential both to the speaker, who needs to know how the audience is reacting, and to the audience, who want to feel involved and needed. It is no good trying to avoid eye contact by looking at the audience's hairlines, as some speakers are taught to do—the effect is shifty, as if the speaker is too unsure to dare to look directly at the people present.

Eye contact should always be brief. There is a tendency for the speaker to want to look too often and too long either at a friend in the audience or at the most important person in sight—the most prestigious client or the managing director. The effect of this is to unnerve the person who is stared at (sometimes causing laughter) and to distract everyone else. As far as possible, eye contact should be indiscriminate—try to look very briefly at as many people as possible, avoiding regular head movement ('the Wimbledon complex') and not forgetting those who are sitting at the extreme left and right on the front row; these are the people most easily missed. In a large hall, it may not be possible to make eye contact with everyone, but it is at least worth looking in the direction of those who disappear into the shadows at the back, so that they know they are not forgotten.

Hands and feet

We are perhaps never more conscious of our own hands and feet than when we are in front of an audience. Our hands take on a life of their own, covering our faces or dismantling anything within reach. Both hands in pockets look too casual for a formal occasion, while folded arms look aggressive and standing to attention looks stiff and uncomfortable.

Similarly, feet can get out of control; frequently they rock back and forwards, or take it in turns to carry the full body weight, sometimes celebrating their time off duty by describing circles or rubbing the other ankle.

Hands are made to hold things, and notes or a pointer in one hand will almost certainly keep that hand quiet. Foot activity is best dealt with by standing with feet slightly apart and the weight evenly distributed—this is also the best posture for good breathing. An additional benefit of the 'presentation uniform' of smart suit and sober shoes is that it persuades the speaker to stand well; students wearing trainers and jeans nearly always stand badly.

Above all, the speaker should move naturally. Standing absolutely still looks and feels odd, and there is usually some acceptable movement during a presentation. The speaker may move across to change a visual aid, for instance, or to allow the next speaker in the group to take over, or to give information by deliberate gesture ('this big'). All this activity holds the audience's attention and gives the speaker the opportunity to change stance in a natural way. Generally speaking, movement is acceptable unless it becomes regular or looks fidgety, and such problems should be picked up at the rehearsal stage.

Every presentation is, to a greater or lesser extent, an exercise in persuasion. The speaker has objectives (see Chapter 1), which usually include convincing the audience that the company is the most suitable, the product is the best on the market, the speaker is thoroughly competent and professional, and so on. The body language has to reinforce these messages. It is no good producing the most brilliant data, convincing evidence and even the most attractive price if the audience cannot be persuaded to believe the speaker. Body language, as we have said, is very powerful, and it must support the words. We can imagine how we would feel if someone expressed deep sympathy while grinning broadly, or warned us of danger in an off-hand manner. Yet presenters may talk about their excellent product or their brilliant research findings and at the same time look desperately unhappy, as if they would make a run for the door at the slightest opportunity.

There are particular moments when it is appropriate to smile at an audience: in greeting them at the start and in thanking them at the end, for instance. In just the same way, positive, beneficial information must be given brightly, with an encouraging and alert facial expression. A strong point may be emphasized by a slight movement towards the audience—a 'we want to co-operate with you' gesture. Imagine (it has happened!) that a speaker says 'We're sure you will want to buy our product' while moving back, away from the very people who should be queuing up to buy!

In group presentations, there is an additional danger that the speaker's own colleagues may look very unhappy (or even shake their heads) while a colleague is speaking. The audience will be very quick to pick this up, and will feel that there is disagreement, or at least boredom, within the presenting team. Each member of the team is on show throughout the whole presentation, and the audience must be convinced that they are seeing a united, supportive and enthusiastic team.

Enthusiasm is, as we have said, infectious. Audiences respond very readily to obvious commitment and excitement on the part of the presenter, and the speaker in turn will respond to the interest and involvement of the audience. But be warned—boredom is equally catching!

Key points

- Non-verbal communication (body language) is specific to each culture; check what is appropriate for an overseas audience
- Eye contact between speaker and audience is essential to both
- Do not use your hands in a fidgety manner (for instance by rattling loose change in your pocket); holding cards or a pointer can help you to keep your hands under control
- Stand with your weight evenly distributed between your feet; keep still but not unnaturally still
- Move decisively when movement is called for, at a change of speaker or visual aid
- The speaker's body language must reinforce the message
- Smile at the audience at the beginning and end of your presentation
- In a group presentation, the whole team is on show and must not distract the audience or look unhappy while a colleague is speaking
- Both enthusiasm and boredom on the part of the speaker will soon be communicated to the audience

NERVES AND CONFIDENCE

The two most important qualities for a successful presenter are nerves and confidence. Gaining confidence is not easy; there are techniques which help, and advice is widely available through courses and through books like this, but the best that such guides can hope for is to encourage the beginner and improve the skills of the experienced practitioner. To give a good, professional presentation (and we firmly believe that most professional people can achieve this), *you have to believe that you can do it.*

Such confidence comes from a variety of sources: knowledge of the subject, thorough preparation, rehearsal above all, and regular practice. Reading about public speaking is never enough (though clearly worth while); it is essential to go through the experience, however brief and informal the first attempt might be. People do survive, and once they see that there is life after the presentation, they feel more secure the second time round; if the performance is praised, confidence grows, and once the speaker realizes that audiences are almost always sympathetic and supportive, it becomes apparent that the ordeal may actually be enjoyed.

After many successful presentations, this initial confidence will have grown until it can be called on whenever the occasion presents itself— though it will always be based on the combination of knowledge, preparation and rehearsal described above. However, the second quality, nerves, still has to be faced, because *the nerves won't go away.* Indeed, it is essential that they shouldn't; a wise Head of Department said to one of the authors, many years ago, 'The day you stop being nervous, retire.'

It is, of course, perfectly natural to be scared. The speaker is going to stand up, alone, perhaps on a platform, and talk to a large group of people for 20 or 30 minutes, and during that time any one of a dozen different disasters may strike. If they do, there is nobody else to help—the speaker will have to cope. 'If the earth does indeed open up,' said an experienced speaker, 'you are the one that gets swallowed.' The speaker is vulnerable throughout the presentation, and of course this results in nerves. The most important thing is to realize that nerves are good—they are one of the biggest assets a speaker can have.

The value of nerves

Nerves are, in a sense, a compliment to the audience, and a good audience recognizes this. If the speaker appears to be casual and over-confident, the audience will feel that they are seen as unimportant, that any collection of people would do. On the whole, audiences quite like a speaker to appear a little tense (they are thinking 'thank goodness it's not me up there!'), because it shows that they, the audience, matter, and that the speaker wants to make a good impression. They have, after all, given time and perhaps money to be there, and they like the speaker to be aware of this. Nerves are an important part of the relationship between speaker and audience.

Nerves also produce a flow of adrenalin. Some people respond well to examination stress, feeling that they give of their best under pressure; similarly, many speakers find an 'edge' to their performance, a brightness and flexibility in their response, just because they are nervous. They tend to look more alert, and there is a sense of excitement in the presentation which would be lacking if they were too much at ease. Oddly enough, this can itself generate a sense of enjoyment; inexperienced speakers sometimes say in surprise, 'I was very scared, but once I got going, I enjoyed it.'

There has to be a balance. It is natural and right to be nervous. If the presentation has been thoroughly prepared and rehearsed, the speaker should also feel confident. Perhaps the best frame of mind just before the presentation is, 'I'm very nervous but I know that's a good thing; I'm also sure that I can do this well.'

Over-confidence

A common cause of stress is, in fact, over-confidence. A speaker may feel, in giving the presentation for the tenth—or thirtieth—time, that it is just another performance, and may therefore not prepare thoroughly or bother to rehearse. Then, in front of the audience, the speaker is suddenly faced with old and illegible notes, uncertainty about the continuing accuracy of one of the facts, or a particularly difficult question—and the nerves take over, together with a profound feeling of embarrassment and discomfort.

The problem can also exist in the opposite form. Extreme nerves can persuade the speaker to put off preparation ('If I don't think about it, perhaps it will go away'), or to avoid practising the presentation, and thus the main sources of confidence, preparation and rehearsal, are ignored. Confidence comes only with knowing both the subject and the audience's needs, and with trying the whole presentation out several times, with visual aids, preferably at least once in front of a sympathetic audience. As time passes and the speaker becomes more experienced, nerves will be less of a problem and confidence will increase, but if confidence takes over completely and the nerves disappear—retire!

Coping with nerves

Nerves can be a great problem for the inexperienced speaker, indeed, dread of being nervous may be worse than the nerves themselves. Nerves affect people in different ways. As we have shown, nervous stress can itself prevent adequate preparation, and this is clearly disastrous. It may be possible to break the ice gradually, with the help of colleagues and friends. The speaker can try, very informally, to talk for a few minutes about a hobby to a group of friends (perhaps sitting down to speak at first); it may be possible to stand on the platform looking at an almost empty hall and then to have a conversation with a few friends at the back—this is surprisingly helpful in giving the speaker a sense of the size of the room and the volume needed without the stress of actual performance. The speaker might join a group for a small in-house presentation, taking a minor role at first. It is always useful for first-time speakers to look at other supportive people from the speaker's position in the room or hall, so that the occasion of the presentation is not the first time they have stood in this exposed place.

The support of other people is essential if the speaker is to come to terms with (not be free from) nervous stress. In order to get such help, it is important to be honest about the problem; pretending that all is well until the last minute is no use at all. At the appropriate stage, at least one rehearsal must be to an audience, even if it consists only of three or four friends or colleagues. The role of this group is crucial: essentially it is supportive and confidence-building, but it must also check that the speaker can be seen and heard, and must give tactful and encouraging advice if things are going wrong. Inexperienced speakers tend to be highly and often unnecessarily self-critical, and their confidence must not be undermined.

Breath control and relaxation

Rehearsal is essential, but as the presentation approaches, the speaker might also want to try relaxation techniques, breathing deeply and allowing the body to relax slowly. Correct breath control is important both for coping with nerves and for the effective use of the voice. Nerves make us all tense up, hunching our shoulders forward and upward; this has the effect of constricting our vocal cords and stiffening our arms (often resulting in clenched fists and clammy hands). In this state, we cannot breathe properly and gasp for air, thus producing a dry mouth and even more tension.

This sad state of affairs has to be reversed, and we should start with our shoulders. Move them freely up and down and allow them to drop to a natural position, at the same time taking one or two deep breaths and slowly releasing them. This routine relaxes us both physically and mentally, so that we breathe more freely and our voices relax to their normal pitch. With correct breathing from the diaphragm, we will start to feel more at ease, and we can produce more volume, if it is needed, simply by putting gentle pressure on the diaphragm.

Some people find it helpful to have something to hold, such as a pen or the overhead projector pointer. As long as the speaker resists the temptation to fiddle (clicking the pen in and out, or constantly extending and contracting the pointer), this is acceptable, and much better than gripping the lectern or the back of a chair so tightly that white knuckles may be visible to the front row of the audience.

But most of all, *believe* that you can present well. The confidence that comes from thorough preparation and rehearsal enables most speakers, however nervous, to give a professional performance. With practice, the perfect balance between nerves and confidence can be achieved.

Key points

- Believe that you can give a good presentation
- Both nerves and confidence are essential to a good speaker
- Confidence comes from knowledge of the subject, thorough preparation and rehearsal. Keep in practice—make presentations whenever you get the chance
- Nerves are an enormous asset in building a good relationship between speaker and audience: they keep you alert and give an 'edge' to the presentation, a sense of excitement which is enjoyable to both speaker and audience
- Over-confidence is as dangerous as under-rehearsal; true confidence grows naturally with practice
- Rehearse in front of friends and colleagues—their comments are helpful and encouraging
- Breathe deeply and allow your shoulders to relax; hold a pen or a pointer if it helps you
- Remember that most professional people can give a professional performance

NOTES

Most speakers, on most occasions, need some kind of prompt during their presentations. The stress is bad enough without the additional dread of forgetting what comes next, and speakers often feel that nerves will make them 'dry up' and forget all that has been rehearsed.

As we have said elsewhere (page 102), thorough preparation and rehearsal provide the best answer to nervous stress, and it is certainly true that if we have rehearsed with care, we are much less likely to forget our message. Nevertheless, it is wise to have notes, even if they remain in the pocket, as the very fact that they are available gives us confidence—and indeed may mean that we do not actually need to look at them.

The form of notes will vary according to the occasion. Presenters at technical conferences, speaking about complex issues to a knowledgeable audience, may choose to have a high percentage of their material written out. At the other extreme, speakers on informal occasions, perhaps to a small group of colleagues, may speak almost 'off the cuff' (although if they are wise, they will have brief notes to guide them and will have tried out what they want to say). In between, notes are almost essential to ensure that the presentation flows smoothly and covers all the information which the speaker intended to convey.

Inexperienced speakers often write out their presentations in full on A4 paper. This presents a whole range of difficulties, not least the temptation to read rather than to talk. Reading at an audience is nearly always disastrous: the speaker has no eye contact, loses any kind of rapport with the audience and gets faster and faster, until people stop listening and are mesmerized by the sheer rush of words—none of which they will remember. Problems are caused by the A4 sheets themselves: they are unwieldy to hold, and bend towards the audience, so that the front row has an exciting time trying to read the words upside down. Paper is also noisy, surprisingly so if the nervous speaker shakes or fumbles several sheets at a time.

Worst of all, if the speaker loses track of the information and actually needs the notes, there they are, hundreds of small typed words close together on a large piece of paper, impossible to identify—and useless as a prompt. Notes, by the nature of their job, must be easy to scan and clearly highlighted.

Using file cards

The most useful notes are key words and phrases written on file cards, with lots of space and colour. File cards are easy to hold and to use; each card can be moved unobtrusively from the front of the pile to the back as the speaker finishes with it. Cards are never noisy, however nervous the speaker, and they are reassuringly firm in the hand. As long as they are numbered, it does not even matter if they are dropped. They are also a discreet form of prompt and do not distract the audience from what is being said.

Ideally, a card should have no more than perhaps 20 words on it, written as key phrases rather than complete sentences. These are easy to read in one quick glance, but there is another less obvious advantage. A sentence is complete when it is written, and is therefore fixed in form. It exists in the same way whether anyone hears it or not, and it is likely to sound 'written'; there is a great difference between the English language as we write it, which is formal, and the English language as we speak it, which is full of abbreviations and much less formal in style. A key word or phrase has to be constructed from the notes into a sentence there and then, in front of the audience—it has no existence as a complete statement until it is created on the spot by the speaker. It is therefore immediate, formed for the specific audience on a particular occasion, and this is how it sounds. It has a life and excitement of its own which listeners will pick up and appreciate. Lecturers who have to repeat a lecture to a second group of students are aware that although the information is the same, the words will be in a slightly different order the second time round, re-thought for a new audience; if this is not so, the second lecture will be lifeless, and the lecturer may start to speed up and lose emphasis, while the students feel that they are somehow being excluded.

For these reasons, the presenter should write notes as key words and phrases on the cards, although there is one exception to this: many people like to write out the first couple of sentences in full as a reassurance that they can make a good start without hesitation or confusion. The fact that these sentences are written out means that they are unlikely to be read—the act of writing will help to fix the introduction in the speaker's mind.

Of the 20 or so words on the card, three or four should be highlighted. The speaker can then glance at the card and instantly see the next few major points and the order in which they come; it takes only a moment to read the whole card. As with most pauses in a presentation, the speaker may feel that the reading time is five minutes long—the audience hardly registers it, or perceives it as five seconds at most.

Some information, by its nature, has to be precise and should always be written in full on the cards. If the speaker quotes from someone else, or gives a reference, it has to be exact; prices, sizes, numbers of all kinds are also in this category. One of the odd effects of nerves is that we can totally forget something which we know very well, and to talk about the benefits of a product and then suddenly to forget its price—or to get it wrong—will do nothing for our credibility. As with the opening sentences, the fact that we have written out this information helps to ensure that we will remember it. (This problem, and its solution, applies also to people's names. It is all too easy to forget the name of a colleague just as you are about to make the introductions, which is embarrassing.)

Some of the information on the cards will be there simply as personal prompts for the speaker. Nerves can persuade us to use a visual aid too early, simply to distract attention from ourselves to the screen, in spite of the fact that we have chosen and rehearsed the optimum moment. If we follow this instinct, we may undermine the impact we have planned. It is a good idea to write a message on the cards, using a different colour from the main text, as a reminder that at *this* point we should show the ohp transparency or start the slides. Timing the visual aids is important; timing the whole presentation is crucial.

The importance of timing

Speaking to an audience has a particular and unnerving effect on the speaker. He or she loses all sense of time. It is almost impossible to keep track of the passage of time, or to calculate how much time is left, while speaking. Yet to overrun the time allotted is the cardinal sin of the presenter. At a conference, you may make life difficult for the next speaker, embarrass the chairperson or even bring about the loss or shortening of the coffee break, none of which will do much for your popularity; in speaking to clients, you may irritate or frustrate the very people you want to impress; worst of all, there is always the danger that the most important or influential member of the audience will walk out. This may be because of a train or plane that has to be caught, but you will not know this. You will imagine that the walk-out was the result of boredom or deep disapproval, and your confidence will plummet. The timing must be right, which means that if you have been asked to speak for 20 minutes, you will ideally speak for 18. Nobody will worry about being spared two minutes ('stop while people still want to listen' is old but good advice), and you have a tiny bit of time in hand in case there is some sort of hold-up, a bulb blowing or a slide getting stuck in the projector.

Help with the timing is available through the notes. A card which says only 'HALFWAY THROUGH', in red, will act as a useful signal to the progress of the talk. A card which reads 'FIVE MINUTES LEFT' will allow the speaker to make a quick assessment of how time and material are passing. If there is time in hand, an extra example might be included; if time is short, some details might have to be left out. (In a recent presentation, one of us heard the speaker say, 'I'm running out of time, so I shall have to speak faster.' This is never an option for a well-prepared presenter.) Part of the planning (see page 127) of a presentation is to decide what can be added or taken out as time dictates. This has to be a decision taken in advance, as it is almost impossible—and certainly highly dangerous—to try to make such a judgement while in full flow. If these decisions are made in advance, the cards can be marked accordingly, and the audience will never know that an adjustment has been made.

Notes at conferences

Earlier in this chapter, we said that it is sometimes necessary to use a fuller script if the information includes a great deal of detailed scientific or technical detail, for instance a paper given at an academic conference. Much of the guidance given above still applies; the text should be printed in large, preferably bold print (never capital letters, which are difficult to read), with wide line spacing and key words highlighted. It may be necessary in this case to use a lectern, although as a general rule we would not recommend such a barrier between speaker and audience—a barrier, moreover, which keeps the speaker rooted to the spot and unable to use much body language. The lectern, if it must be used, should be at the correct height, and the light sufficiently strong; the speaker who has to peer at the script will soon lose contact with the audience.

Visual aids as prompts

A different but extremely useful form of notes is provided by the visual aids (see page 116). If the speaker is explaining a diagram which is presented on a slide or a transparency, it is itself the notes; the speaker should be able to take the audience through the process, for instance, without needing any other prompt. It goes without saying that the presenter must be totally at ease with the diagram and able to explain it in a clear, logical way.

Many speakers need practice at using cards or any other form of notes, and in the process of rehearsing, they become more familiar with the material and so less and less in need of notes. They may even reach the point at which no prompts are needed—which is the precise moment at which over-confidence may become a danger. True confidence comes from knowing that the notes are within reach, in the unlikely event that they are needed.

Key points

- Notes are essential to ensure that you say all that you intended to say and that the presentation flows smoothly
- Use notes that are appropriate to the occasion, but do not write out the whole text on A4 paper
- Notes must be easy to scan and clearly highlighted for quick reference
- Use file cards with key points (not sentences) and do not write too much on a card
- Remember that written English is much more formal than spoken English, and make sure that you talk to the audience
- If you are very nervous, write out the first couple of sentences in full, for reassurance
- Precise information, such as prices or measurements, should always be on your cards

- Write messages for yourself about visual aids or timing, in a different colour
- If you feel that you need fuller notes because of your highly technical subject matter, space and highlighting will still be important—try not to read at the audience
- Your visual aids may themselves be a useful prompt, but be sure that you are at ease with them and totally familiar with their content

GROUP PRESENTATIONS

A great many—perhaps most—industrial presentations are given by groups rather than by individuals. In-company seminars, presentations to prospective or existing clients and customers, project team briefings, for instance, may well be given by teams of three or four of the personnel involved. Within the academic community, undergraduate project presentations are frequently given by pairs or small groups of students working together, and presentations to sponsoring bodies are often given by a research team.

There are great advantages in presenting as a group. The company is able to display a range of expertise, and the client sees the particular skills of each member of the project team. Variety within the presentation is easy to achieve, each change of speaker allowing the audience a brief break and a new surge of concentration.

Moreover, a group can show valuable qualities which are difficult to assess in any other way. Teamwork is often critical to the job, and a potential client will look for a display of organizational skills, flexibility, and team loyalty and support. 'If they can't get their act together for the presentation, how can we trust them to work together on site?' is a natural and valid reaction. Fortunately teamwork, like the other qualities, can be rehearsed—always assuming that the team comes together before the date of the presentation.

Rehearsing as a group

Rehearsals take time, and time is money. However, expecting a team to manage an effective presentation on the spur of the moment is a false economy; the group must be allowed to go through the complete presentation at least three times, preferably at intervals over several days. Individual rehearsal is valuable, but the sum of the separate parts of the presentation is nearly always greater than each part would suggest; in this respect, three 10-minute presentations add up to about 35 minutes. As we have said (see page 108), accurate timing is essential, and the team cannot be sure how long their talk will last until it has been tried out as a whole, including the visual aids. It is surprising how much time is involved as each speaker collects the visual aids and moves out of the way to make room for the following contributor. No individual should have to adjust what is said on the spur of the moment because someone else has overrun, and the last speaker should have the same opportunity to influence the audience as the first, though not necessarily the same length of time—the subject matter will dictate how much of the time available is allocated to each speaker.

Careful rehearsal will result in accurate timing, and also in a sense of trust between speakers. In a group presentation, each presenter carries responsibility not only for the individual section but also for group support and co-ordination. A true horror story concerns the company which added a director to the presentation team at the last minute, in order to impress the client. The team, which had done its own rehearsal, was not impressed, and even less so when the director started the introductions and then found that he did not know the name of one of his colleagues. So much for group co-ordination!

Number of speakers

There is no rule about who should speak, or how many speakers there should be. The audience will sometimes dictate this, for instance a prospective client might wish to meet the whole project team or might ask that the project manager be present. Otherwise, the choice of speakers depends on the range of expertise which the company wants to show, and the number of speakers depends on the time available. It is generally unwise for anyone to speak for less than five minutes, apart from a chairperson introducing the team, as the effect can be fragmented and distracting. Allowing a single speaker to continue for more than 20 minutes puts a strain on that speaker, and may make the audience wonder how long everyone else will speak for and how late it will be when the event is concluded. A change of speaker adds variety and so aids concentration.

Chairing the group

The group may choose to have a chairperson or co-ordinator, who introduces himself or herself and the others. Everyone should be introduced, including colleagues who are not taking part in the presentation but whose particular expertise is needed for answering questions. The chairperson will also be responsible both for receiving questions and allocating them to appropriate speakers (see also page 127, handling questions) and giving a focal point to the whole presentation, which must appear as one event rather than as a series of disparate talks.

For the same reason, there should be a logical overall structure to the team presentation. Each individual contribution needs its own structure (see above), but the general pattern should be made clear by the chairperson or the first speaker. A group of construction students, making a presentation to fellow students about the design of their building from the point of view of a student in a wheelchair, began like this:

> We're going to describe the problems faced each day by a student who is confined to a wheelchair. First we shall show the difficulties of getting into the building, up the ramp and through the outer doors. Then we shall look at mobility around the building, the corridors and the lifts, and finally we shall look at the problems presented by specific rooms, such as opening doors and windows, or using the coffee machines in the common room.

The logical pattern is clear, and the audience also knows what is excluded—use of the university library, for instance.

Group image

The group must present a coherent image to its audience, as this is one of the ways of showing the efficient organization and management of the company. A consistent style must be agreed, for instance in the form of address used in the introductions (first name and surname is the most common form for men and women nowadays) and when the team members refer to one another. This style extends to the formality or informality with which the presentation is given; if one member of the team adopts a very casual and familiar style when the others are very formal and restrained, the effect can be slightly comic.

The same consistency applies to the visual aids. It is possible to have all the slides, for example, made together, so that they all have a similar border or the company logo in the corner. Any material given out to members of the audience can then have the same design, and a pleasing effect of control and attention to detail is achieved.

The physical organization of the group within the room is important. The whole team should be visible to the audience, but those who are not speaking should not obstruct the view of the speaker and the visual aids. They should also not distract from the speaker by whispering to one another or ostentatiously looking at their notes. If they are not paying attention to what is going on, they are implicitly giving the audience permission to lose concentration too; what is worse, they will not pick up any problem which faces the speaker. Nerves can make one member of the group say something planned for a later section, or leave out an important detail. If the next speaker is alert to this, it is often possible to readjust the material so that nobody in the audience realizes what is going on.

Waiting for one's turn to speak is never easy: in full view of the audience and yet with nothing to do, the next speaker hardly knows where to look or how to give a good impression. Sitting upright but leaning lightly against the back of the chair is usually effective; looking approvingly at the speaker and from time to time at the audience to pick up any non-verbal messages gives a positive signal to others. Avoid sitting in a straight line confronting the audience, which is most uncomfortable—if the chairs are placed at an angle, everyone will feel more at ease.

The position of visual aid equipment needs to be planned, so that no member of the group is likely to trip over a flex or indeed over a colleague— a predicament which is not unknown. It should be easy for the speaker to move to the side and sit down and for the following speaker to move into position.

Taking a small part in a group presentation is a good way of gaining confidence for the inexperienced speaker, and there are several jobs which are essential to the smooth running of the whole presentation but which require little actual speaking. Someone has to co-ordinate the visual aids, to dim the lights and to put them up again, to give out brochures or other handout material, to check that the equipment is turned on and correctly focused, and to make sure that the team's chairs are well positioned. These are all trivial activities in themselves, but careful attention to detail is what makes a presentation professional—and therefore successful.

Key points

- Use the many advantages of presenting as a team, remembering that both responsibilities and stress will be shared
- Your group must rehearse together, not least in order to get the timing right
- Allocate the time on the basis of the aspect to be covered, allowing each speaker at least five minutes; the number of speakers may depend on the range of expertise available
- A change of speaker adds variety and increases the audience's concentration
- All members of your group must be introduced to the audience, whether or not they are going to take part in the presentation
- Your group must present a coherent image, in style, appearance and visual aids
- Support each speaker in your group by showing interest and commitment—and watch out for any potential problems
- Check the physical organization of your group within the space available
- Each of you will depend on the others—a presentation given by a group will be judged on a group basis
- The quality of your teamwork in the presentation will be seen as indicative of your teamwork on site

PREPARING THE VENUE

The place where the presentation is to be given requires a good deal of thought and preparation. Obviously there are advantages in using home ground: the size, shape and facilities of the room are not only known but familiar, and, most important of all, rehearsals can take place there to give a sense of security and confidence to the speakers. Booking the room not only for the event but also for rehearsals should always be an early consideration.

Seating arrangements

The size of the room is important. It must be sufficiently large for comfort, but not so big that the audience feels lost. It is almost as bad to be one of an audience of 6 in a room designed for 100 as it is to be one of 40 people in a room in which 20 is a sensible maximum. People need their own space, which includes the possibility of moving a chair slightly to get a better view and having sufficient elbow room to make notes, but they may also feel uncomfortable if they are surrounded by rows of empty seats.

There are odd national instincts about where people choose to sit. Unlike some of their European neighbours, for instance the Germans, a British audience will fill the room from the back, leaving the embarrassment of the front seats to latecomers (not a very wise procedure from the point of view of the speaker, who needs as little disruption as possible!). The speaker may need to take control, preferably before the audience arrives, by physically removing some of the seats at the back, or by putting a handout on the seats at the front to encourage the audience to sit where they may easily be seen and addressed.

The 'examination room' look should be avoided at all costs. The sight of rows of people sitting at an equal distance from each other will bring back horrible memories to both speaker and listeners, unnerving both. Straight lines should also be avoided if possible—even the simple turning of the end two or three seats to give the impression of a wide horseshoe is better than direct confrontation between a line of speakers and a line of listeners.

If numbers are small, there is more flexibility. A dozen chairs can be set out in a semicircle, with the added advantage that most people can see most other people: discussion will then be freer than if a speaker from the audience has to talk through the back of other people's heads. This layout may be more convenient than seating the same number round a table, which is often seen as a good way of getting an audience to contribute but which has the disadvantage of making eye contact between people on the same side of the table almost impossible.

There are of course other considerations, for instance whether latecomers may need to join the group as unobtrusively as possible or whether those present are likely to make extensive notes and so need table or desk space. At the other extreme, a presentation to a large number of people will almost certainly be formal in its very nature, and questions rather than discussion will probably come at the end; members of the audience will not expect to make eye contact with one another very often, and it is more important that they should be able to see any visual aids clearly.

Visual aids and equipment

For both audience and speaker, the position of equipment and the visibility of the screen are critically important. The speaker must be able to move freely without fear of tripping over flexes or being hidden behind projectors. Indeed, an essential aspect of planning is the position of speaker or speakers: the organization of a group presentation is discussed elsewhere in this chapter, but it is worth repeating that speakers should not fall over one another or tread on each others' feet—a much more common occurrence than one might imagine!

The visual aids to be used will be planned well in advance of the presentation (see page 118), and the placing of the necessary equipment will be one of the considerations in the choice. In a small room with a restricted audience, too many forms of visual aid can be a distraction; in a large room or hall, some forms of visual aid, for instance the flipchart, will automatically rule themselves out. Both parts of the equipment need to be placed with care—the audience must be able to see the screen clearly and the speaker must, for instance, be able to reach the overhead projector easily to change acetates, or to use a remote control without getting tied up in flex. As in so many aspects of making a presentation, rehearsal is essential not least so that the speaker is familiar with the positioning and the use of the equipment.

Audience comfort

Other 'comfort' considerations are important. Do the speakers need to dim the lights or to use blackout? Would it be sensible to open windows or operate the air conditioning before the audience arrives? Can the speakers check the time without being seen to refer to watches—a well-situated clock provides most welcome guidance to the speaker! Legal and safety requirements, such as exit signs in a lecture hall, should be checked. Nothing in a presentation should be left to chance.

All such planning assumes that the room for the presentation is at the speakers' premises and that they have the opportunity to organize the details. This is, of course, not always the case.

Clients' premises

If the presentation is at the client's premises, most of the decisions will be made by other people. It is, however, perfectly reasonable to ask some questions in advance, such as approximately how many people will be present and what visual aid equipment will be available. Some equipment, such as a portable overhead projector, can be taken by the speakers, but it is as well to ask whether blackout is possible if slides are to be used. Indeed, such a request for information shows forethought and attention to detail, which must surely impress the client!

Key points

Own premises

- Choose the venue with the audience and the visual aids in mind; the size of room should be appropriate, neither too big nor too small
- Plan the seating in a layout which will not intimidate the audience; if discussion is important, try to allow maximum eye contact among the group
- Decide where the speaker(s) will stand and the position of equipment to ensure that speaker(s) and audience can see one another
- Ensure that the audience can all see the visual aids
- Consider the convenience and comfort of the audience, adjusting the lights, ventilation, etc. as appropriate; check any legal/safety requirements
- Rehearse in the venue, using all the appropriate equipment until you are at ease with it

Client's premises

- Ask appropriate questions in advance, especially with regard to numbers and visual aid equipment
- Take your own equipment with you if it seems wise/tactful to do so

VISUAL AIDS

Visual aids should, as their name suggests, provide support for both speaker and listener. For the former, they can be something of a respite, deflecting attention from the individual and onto the screen; for a nervous and inexperienced performer, it can be a great relief to realize that the audience's attention is momentarily elsewhere. They can also provide a form of notes for the speaker (see page 109), and indeed can make possible a gradual analysis of a complex diagram. They make far more impact than words alone can do, and are more easily remembered. Few speakers would want to talk for long without the backup of visual aids.

Audiences love a good, well-used visual aid. It gives them a break from the arduous task of listening, allows them to use their eyes rather than their ears, and, because human beings are attracted by and remember patterns, their job is made easier and more enjoyable. Few audiences would want to listen for long without the backup of visual aids.

It is, however, primarily the speaker that they have come to hear. After all, they could have received a pack of illustrations before they left home, and perhaps not bothered to come to the presentation at all. Visual aids must never be allowed to replace the speaker; their function is to support, clarify and add variety to what is said, but they are essentially subordinate and must remain so.

Quality

Nowadays, audiences are sophisticated and want visual aids which are of the same quality as the rest of the presentation. The 'home movie' effect is disastrous, but still surprisingly prevalent. Handwritten overhead transparencies are no longer acceptable—even undergraduates, with limited resources, are expected to produce their transparencies on a computer.

It is not difficult, though it may be costly, to have good visual aids prepared; most organizations have their own support staff with the necessary computer equipment, and others have access to commercial companies which will work closely with the presenters to produce appropriate material.

This book will not, therefore, describe the process of making visual aids. It is much more concerned with their effective use, and will discuss a selection of the most common visual aids in the light of their advantages and the pitfalls associated with them. There is inevitably some duplication: obscuring the audience's view by standing in front of the screen is a trap for users of both the overhead and 35mm slide projectors—and a trap into which a surprising number of speakers fall.

When to use visual aids

When will visual aids be appropriate in my presentation? This is the first question to ask, and the answer will depend on both the subject and the audience. Some subjects, by their nature, call for particular visual material: an architect describing the design of a building must show it to his clients; a student engineer will show a circuit diagram in order to talk through it and show an appropriate understanding. Visual aids may be used to show movement, such as the flow of air through a space, or a product, in both technical makeup and appearance; they may present a contrast, for instance before and after the installation of a new lighting system. There will be times in almost any presentation when the audience needs to see what is being described.

Exactly what is shown depends to a certain extent on the knowledge and experience which the particular audience is bringing to the event. A group of civil engineers will be less interested in a slide showing the appearance of a new bridge than in its technical make-up—though they may be helped initially.by seeing what the general appearance of the bridge is like, before the details are discussed. This need for a visual introduction should be remembered: it is helpful for the designer to say 'this is a plan of the building as a whole' before 'now let's look at the fire protection we've envisaged.'

Visual aids are chosen in the light of the content of the presentation; this seems obvious, and yet presentations are sometimes built round the available visual aids to the extent that the visuals take over. This is especially true in the case of presenters whose visual aid material is particularly striking—architects with superb drawings and photographs are particularly prone to be led by their visual material rather than supported by it.

Number and type of visual aids

The number of visual aids that can be used effectively in a given time is difficult to assess: it is obviously dependent on the information to be given. Nevertheless, the subordinate nature of visual aids must be borne in mind, and perhaps as a rough guide a maximum of half a dozen overhead transparencies or 10 slides for every 15 minutes is a sensible guide.

Every visual aid takes time. It cannot simply be presented to the audience and then immediately whipped away; it has to be put in position, checked briefly by the presenter for focus and accurate placing (it should not, for instance, appear two-thirds on the screen and one-third on the ceiling or the wall below the screen), and then left until the audience has had a chance to assimilate the information. During this short time, the speaker must remain silent. If the talking continues through the presentation of the slide, the audience will be confused—is it supposed to be looking or listening? Being unsure, it will do neither efficiently. Only when the audience has looked at the visual aid and understood what it is seeing should the speaker attempt to discuss what is being shown.

Visual aids are intended to be used—they are an integral part of what is going on. The speaker should never simply show the picture to the audience and pass on: it must be described and discussed, so that the audience knows exactly why it has been used. All this takes time, and, if there are too many visual aids, there will not be sufficient time for the speaker to make an appropriate impact.

The type of visual aid has to be chosen in the light of the material itself, the audience, and the equipment available. Generally speaking, slides are convenient for showing either general views or detailed diagrams to almost any size of audience, while a flipchart, for instance, is informal in itself and sensible only for a small group (it is particularly helpful at meetings as a way of generating discussion—see page 137). Display material can provide a useful backup to the talk, but it is unlikely to be visible to the audience during the presentation. Handout or brochure material provides a lasting record of the event, and also allows each member of the audience to have access to information which it is difficult to put across in any other way, for example a detailed breakdown of costs, or mathematical calculations.

Too many different forms of visual aid cause confusion: flexes are tangled up, there are not enough sockets or they are in the wrong place, equipment has to be moved out of the way or into position and then refocused. The speaker loses control and the audience feels that the event is turning into a display of technology. One or two types of aid, perhaps with written backup, are sufficient for most purposes, provided that the quality is good and the speaker is clearly at home with both the equipment and the visual material itself.

The most common form of visual aid, all in all, is probably the overhead projector, followed closely by the 35 mm equivalent, but more recent developments such as the liquid crystal display (LCD) projection panel are gaining in popularity. The selection of visual aids depends on a variety of factors which the speaker must bear in mind.

Selection of visual aids

Perhaps the most versatile of all visual aids is the **overhead projector** (ohp). Transparencies can be prepared in advance and used for almost any size of audience, although definition may be lost in a very much enlarged image. The machine itself is easy to carry; the acetates are simple to produce on a wordprocessor or photocopier, and there are few restrictions about lighting—almost any place not in direct sunlight is possible. The most common mistake is copying from a printed text onto acetate in spite of the fact that the lettering is much too small. If the audience is sitting 15 m away, each individual letter should be at least 15 mm high; under any circumstances, letters should be a minimum of 5 mm in height. Only a limited number of words fit comfortably on an overhead transparency; the key message should be towards the top and in the centre, while the bottom quarter or so of the transparency should be left blank—it tends to 'disappear' behind the projector itself.

Overhead transparencies should not give an impression of clutter. Only information which the audience needs to see should be presented, and, generally speaking, numbering and punctuation are unnecessary. Colours should be used with care: black is always dominant, while dark blue, brown and purple project well. Red is good for highlighting, but not for words; green tends to look washed out, and orange and yellow are virtually useless.

Pictures in colour can be photocopied on to acetate to give a professional impact. A corporate image can be incorporated by the addition of a logo to each transparency, and the information can have a 'frame' to improve the layout. Plastic mounts are useful if the transparencies have to be carried about, and these can usually be filed in a ring binder for protection.

Similarly, **35 mm slides** can be made in advance and can be projected in a large or a small room. They enable a speaker to show a very wide range of material, from pictures of sites in any part of the world to complex diagrams, and are therefore very popular as visual aids. However, blackout is essential and lights need to be dimmed, and there is the risk that the audience will go to sleep in the warm, dark room! A common mistake is to use slides which are blurred, too dark or which do not show exactly what the speaker intended. It is not usually appropriate to show only two or three slides—too much reorganization of the room and the lighting has taken place for too few slides—but as with most visual aids, it is not sensible to show so many that they seem to take over the presentation, so that it becomes little more than a slide show.

The restrictions about colour suggested for overhead transparencies tend to apply also to slides, for instance, words should not be shown in red. However, a coloured background can look effective as long as the colour combination is chosen wisely. White or yellow lettering on a blue background is clearly seen and pleasant to look at. Pictures or diagrams presented in slide form should have clear, uncluttered detail, and the speaker will probably use a pointer to indicate to the audience exactly what he or she is referring to.

A small mark in the top right-hand corner of the front of each mount will show the correct position for slides when they are placed in the carousel.

One of the most informal visual aids is the **flipchart**, which is especially useful for a small audience when some interaction is required. It can be prepared in advance or the speaker can draw an illustration in front of the audience—provided that this does not result in the most common problem, a spelling mistake or a drawing which goes wrong! A flipchart is useful at meetings, but is not appropriate in a very formal setting, as it lacks a corporate image and usually carries a handwritten message.

Video is increasingly popular as a visual aid. It is expensive to produce well, but can extend the contributions—for instance, different speakers and moving images can be shown. Potential clients and customers can be given a video as a permanent record and therefore lasting publicity for the company. In spite of all these possibilities, video should be used with care: it can go out of date, and it can overwhelm the speakers whom the audience wanted to hear. If a short video presentation is appropriate (usually to show movement, for instance how a robot works or to illustrate the movement of air through a space), it is often best to show it at the end of the presentation; if the video is shown earlier, it can be difficult to recapture the audience's attention.

Demonstrations are usually well received, as an audience likes to see something happening as a change from just listening. However, there is usually a visibility problem if the audience is more than a dozen or so people, and it can result in a technical presentation becoming a bit like a sales talk. However, the major worry is obvious—a demonstration can go wrong in front of the audience.

LCD panels or tablets are becoming increasingly popular as a means of conveying detailed technical information to a small audience. They are particularly useful in discussion, as the image can rapidly be adapted or developed in line with discussion or questions. Laptop computers and suitable software are needed, and some technical or secretarial support is helpful. As with the demonstration, things can go wrong and speakers may be left without their main tool. On an important occasion, it is worth having a backup, for instance ohp transparencies, just in case there is a problem. Nevertheless, the LCD panel has great potential in small group work.

Handouts remain with the audience as reminders; they can be prepared and checked in advance, and can present an impressive corporate image. Complex material, or detailed costing, is usefully given in this way. Handouts may be given to the audience at the beginning of a presentation, in which case the speaker can refer to them and use them during the talk, or at the end of the event, as confirmation of what has been said. They should never be handed out during a presentation, as they distract the audience (and some people might not be able to see a copy at the appropriate time). Perhaps the most common mistake is to give too much material as handouts, so that the audience cannot find its way round all the documentation—or becomes too absorbed in it to listen.

Visual aid presentation

In using any of these visual aids, it is essential that the audience can see clearly. Neither the speaker nor the equipment should block the audience's view, and all the words and diagrams presented should be easy to read. In a group presentation, the visual material should be co-ordinated so that there is uniformity of style and of colour (for example, if the walkway is marked in green on one transparency or slide, it should be shown in green throughout the presentation). The overall impact of the visual aids reflects the professionalism of the speaker, who will gain confidence from having good quality visual material to show.

Visual aids should be subjected to the most careful checking, both for clarity and for accuracy. They must show clearly what they are meant to show (and as little else as possible), and they must not contain inconsistencies or errors of technical content or spelling. As with any checking (see page 31), it is helpful to have a second opinion. Two real-life examples will show the awful effect of omitting this stage: a group of undergraduates using visual aids in describing their project introduced one of their slides by saying 'This is a picture of the renovated barn we visited. Sorry about the line of trees in front of it.' And at a company open day, one of the directors proudly described the quality management scheme which had recently been introduced. He had key words on a beautifully produced slide. Unfortunately, in the word 'quality', the 'u' was missing.

Key points

- Visual aids are a means of support for both speaker and audience; do not allow them to take over the whole presentation
- The quality of the visual aids must be as professional as the rest of the presentation
- Remember the advantages of the general introductory visual, and then decide what else the audience will need to see
- Do not try to use too many visual aids
- Do not rush your visual aids—give the audience plenty of time to assimilate what they are seeing
- Always integrate the visual material into the talk and tell the audience why they are seeing it
- Choose just one or two appropriate types of visual aid; you must always remain in control
- Some very complex or detailed material is better supplied in handouts: a breakdown of costs, for instance
- Check that all the audience can see and read all the visual aids
- Check visual aids for consistency, technical accuracy, grammar and spelling

7

Spoken Presentation

The formal presentation • Speaking at conferences • Speaking at meetings • Videoconferencing • Speaking on the telephone

An important aspect of the work of most professional people is speaking to an audience; it is often a task which causes more stress than the technical work itself. There are good reasons for this: a speaker is isolated and vulnerable, and, just as the impact of a presentation may be considerable, the price of failure may be high.

Yet formal presentations have become increasingly important: between companies, speaking to actual or potential clients, tendering for contracts, at briefing meetings; within companies, in reporting to senior management or to colleagues, at meetings and perhaps at company open days; outside the ordinary working day, perhaps in speaking at a professional institution, to young or would-be engineers, at public meetings; in applying for jobs, since the personal presentation is often used in appointment or promotion interviews; in education, since university departments of engineering, construction management and architecture, for instance, are more and more aware of the need for students not only to acquire technical expertise but also to be able to communicate it. On all these and other occasions, it is seen as part of professional duty to present well.

Chapter 6 looked at some of the techniques which can be employed in speaking effectively; this chapter puts these into context by considering a range of occasions on which such ideas may be put into practice. Needless to say, a first-time speaker cannot hope to use every good technique at once. It is helpful to take one or two key ideas, such as eye contact and speaking slowly, and to concentrate on getting these right. Gradually, with practice, the speaker will find that they are happening instinctively, and that other techniques can also be employed as confidence increases.

THE FORMAL PRESENTATION

(See also Chapter 8 for a detailed example.)

Audience and objectives

The first stage of preparation for a formal presentation is similar to that for a report (see page 50): it concerns the audience and the objectives. The nature of the audience must be identified in terms of expertise, experience and seniority, by the answers to the following questions:

- Who will be in the audience?
- What do they already know about the subject—are they at the same level of knowledge or will they vary?
- What experience of this type of work have they had? Again, do they vary?
- What level of technical language is it safe to assume?
- Do all members of the audience share the first language of the speaker? If not, how fluent are they?
- How many of them are there likely to be?
- How senior will they be?

The last three questions are much more important in a presentation than in a report. Of course any writer would like to know the first language of the readers, but if some are of a different nationality, it is likely that they will be able to cope, especially with technical words (the enormous advantage in this respect of those who have English as their first language is obvious), or they will be able to get a translation. The spoken language is always more difficult, and the considerate speaker will bear that in mind.

Reports, of course, can be sent to an unlimited number of people, world-wide; a presentation has to be prepared for a limited number and will be very different depending on whether that number is 5 or 50 or 500. The speaker needs to know before starting work. The seniority of the audience does not always matter, but it may be a guide to the way in which the material should be presented and to the terms used (much training takes place on courses; if it is for senior people, it will take place in a seminar; very senior people, who should know anyway, have discussions on the subject).

Two sets of objectives have to be considered: the speaker's and the audience's. They are rarely simple. The speaker may have both an immediate and a long-term objective—for instance to show the company's expertise in a particular area, in order that a potential client may look favourably on that company when a new project is in prospect. Other unspoken objectives may be important: the speaker may want to impress the client in order to ensure a position in the project team, and his or her own senior managers to improve promotion prospects; a personal point of view may be reinforced or accepted, and professional rivalry may make the speaker anxious that some other person or company does not win the desired contract. All presentations are to a certain extent opportunities for 'selling' the individual, the profession, the company, an idea, and the wise speaker keeps this in mind throughout the preparation.

Members of the audience will also have objectives—after all, they are giving up valuable time in order to attend the event. What will they get out of it? They may be better able to make a decision, for instance about whom to employ; they may learn about new areas of expertise, products or processes; they may have their prejudices removed or reinforced, and to a certain extent they will also be 'selling' themselves—'we are a first-rate company and we want to choose first-rate people to work with us, people who will be proud of a link with us.' Again, motives may be very mixed.

For the most part, there is a matching of objectives: they want to buy what we want to sell; there may, however, be a conflict: they have already decided to employ someone else and are simply fulfilling obligations in listening to us. It is not possible to achieve both sets of objectives on such an occasion, but there will be other times, and the effect of a good, professional presentation may be lasting.

The speaker must identify all these objectives as fully as possible, in order to assess what the audience will want to get out of the occasion. This is a stage which is often skimped, the assumption being that what is needed is information. This is, of course, true, and if the facts are not given or are not correct, then the result is likely to be disastrous for the speaker. Well-chosen, accurate material must be taken for granted, but it could be supplied by post or fax, so that the expense of the presentation might be spared. The audience needs to meet the speaker in person for other reasons apart from acquiring facts.

Questions may be asked and answered by fax, but the personal exchange of question and answer is particularly revealing—*how* the question is answered may be just as important as the answer itself. Questions may give rise to further, more searching questions, and the speaker will be watched carefully for signs of doubt or unease. Perhaps the most efficient way of finding out exactly how much a speaker really knows about the subject is to ask probing questions and to watch as well as listen to the response. A speaker cannot be expected to know everything, but the way a 'don't know' is handled can be very revealing (see page 127).

The need for professionalism

Above all, the speaker must be seen to be professional. It is not easy to define this word, but it comes from a series of impressions, not least the initial ones. A smart businesslike appearance makes an immediate impact, and indeed correct dress has an effect on the speaker as well as the audience. Student presenters often rehearse in the usual student dress of T-shirt, jeans and trainers, which is fair enough. They stand awkwardly and appear too casual and ill at ease. When they make an effort to look professional at the presentation itself, wearing suits, plain shirts, ties if appropriate, and particularly proper shoes, they are amazed at the difference in themselves. They stand well (weight evenly distributed) and breathe correctly; they both look and feel more confident, and the audience immediately recognizes a greater professionalism.

Confidence is impressive (it has, of course, nothing to do with the absence of nerves—speakers need to be nervous; see page 102), and it transmits itself quickly to the audience. The message is 'I clearly believe in what I am saying and so you can, too'. It encourages trust between speaker and listener, and trust is essential if the presentation is to be successful. The speaker must build a rapport with the audience, looking at them, smiling at them when appropriate, being responsive to them throughout. This is part of successful 'people chemistry', the indefinable ingredient which makes people respond favourably to a speaker; it is built up quickly but can be lost just as quickly, for instance by aggression or discourtesy on the part of the speaker. Without this rapport, the presentation will be flat and uninspiring; when it is evident, the audience is alert, interested, pleased to be listening, on the speaker's side. It is often this quality which makes an employer choose one candidate at interview rather than several others who are equally well qualified and experienced.

Other impressions are also important. If a problem arises, is it handled speedily and competently? In a group presentation, is there evidence of teamwork? Are the group members courteous and encouraging to one another? Is the team well managed? Above all, is the speaker enthusiastic about the topic, committed to the point of view being propounded? Enthusiasm always attracts an audience—if the speaker cares about the subject, so will they. Obviously it is possible to overdo this enthusiasm, but if it is not there, the listeners will wonder if the speaker would be as casual and uncommitted at work as in the presentation, and this suspicion will undermine the speaker's credibility.

Some, if not all, of these considerations will be in the minds of the audience as they watch and listen; they are too often ignored in the preparation. The need to rehearse is obvious: only with thorough preparation and practice will the speaker be sufficiently at ease with the material and style of the presentation to generate the required confidence, commitment and professionalism.

Presentation style

The style will vary with the size of the audience and the personal approach of the speaker. At a formal presentation, the speaker will almost certainly be standing, both as an indication of the formality of the occasion and in order to see and be seen by everybody present. Some speakers are naturally reserved and prefer some kind of barrier (lectern, table or whatever) between themselves and the audience; others are by nature informal, and want to be as close to the audience as is sensible, in order to establish a warm sense of two-way communication. To a large extent, speakers must know their own particular approach, although in the case of a formal presentation, it is dangerous to be too informal, while too great a formality may make the speaker appear remote and perhaps rather cold. Colleagues and friends will usually be able to comment on this at rehearsal.

Preparing the information

In the light of the audience and the objectives, the presenter will gather the appropriate material, making sure that there is more than could possibly be used in the time, as the sense of reserves of information communicates itself to the audience. Facts will be checked, opinions discussed and company policy ascertained. Nothing will be left to chance, least of all the answers to questions.

Most speakers have a particular dread of the question and answer session. They usually say something like 'I know what I'm going to say, and that will be all right, but I dread the questions. What are they going to ask?' This is precisely the question which the presenter must ask early on in the preparation. 'If I say this, what might they ask in response?' An essential part of rehearsal is to sit down with a small group of colleagues and say, 'Now you've heard my presentation, what would you expect the audience to ask?'

If this stage is handled carefully, at least 75 per cent of the questions can be predicted, and the answers prepared in advance. Questions in presentations fall into two categories: those which the speaker should be able to handle, and to which a 'don't know' reply is embarrassing, and those which are outside the speaker's frame of reference, or obscure, or dependent on other specialist knowledge, and these the speaker may reasonably not be able to answer. There are various ways of handling this latter situation, of which 'I'm afraid I don't know [*or* don't have that information with me] but I will check with my colleagues and ring you within the next 48 hours' is a generally accepted norm, provided of course that the speaker fulfils the promise. It may also occasionally be worth asking the audience for help, or suggesting a discussion with the questioner at the end of the session. It is never worth bluffing: the speaker who resorts to this will undoubtedly find that the world expert on the topic is there in the audience and only too eager to put the speaker right.

It is worth in this context mentioning the situation of student presenters. They are almost certainly in the unusual and rather strange position of telling people what they, the audience, already know. It is helpful if the listeners are a mixture of fellow students and staff, so that some at least are learning from the event, but the key people, the staff who will mark the presentation, may well be experts on the topic chosen. The level of the presentation must be agreed in advance, and the questions should then be chosen on that basis; the students themselves will almost certainly get credit for knowing the answers or, if the question is beyond them, for handling the situation with tact and courtesy.

Structure

The material for the presentation, when it is collected, must be sorted and organized in a logical way, perhaps by the use of a technique such as the spider diagram mentioned earlier (page 13). In the case of the presentation, it is often a good idea to collect far too much information, sort it, and then select in the light of the particular audience and the time available. Some of the extra material may well be useful in answering questions. Most presentations can then be given a structure, on the following basis:

- Introduction, including names and perhaps credentials of speakers, area of expertise, subject, sequence in which the subject will be handled. This section should not be rushed, as it establishes the rapport between speaker and audience
- Main ideas, in a logical order, with appropriate visual aids
- Conclusion, in which the most important information is repeated and highlighted
- Thanks—the audience is briefly thanked and questions are called for

Each of these sections will now be discussed in more detail.

Introduction   The introduction is more important than is often recognized. The subject may have been decided and the speaker named before the event, and it seems to be a waste of time to go through this again. But the audience needs reassurance, especially if they have not previously met the speaker, and they will also want to know how the topic is going to be handled. It is one thing to sit and wait for the section which is of particular interest and know that you will have to listen to other aspects first; it is quite another to sit and wonder whether your special interest will ever be dealt with or whether you have wasted your time in coming. In a group presentation, each member of the group must be identified, with the specialist knowledge of each made clear. The audience wants to know who these people are and why each is speaking about the particular aspect. (See also page 110 for the structure of group presentations.) They will also want to know how long the presentation will last, and when they may ask questions.

All this preamble has a secondary purpose. When an audience starts to listen to a new speaker, they are making all kinds of rapid assessments—can they see, can they hear, would it help to move their chairs slightly, does this person look like what they expected/sufficiently professional ... and, on a more personal level, what the speaker's accent is or why he should be wearing such a very loud tie. These may be only semi-relevant to the occasion, but most people take an interest in other people and sum them up at the first meeting. The speaker must allow for this. If the very first sentences contain complex technical data, the audience's mind will not be attuned to what is said, and important, perhaps essential, information is lost. It is much safer to spend a very brief time going through the introductory material so that by the time the speaker reaches the meat of the session, the audience is giving 100 per cent attention.

During this introduction, the logical ordering of what is to follow will be given—'first I will look at, then I should like to discuss, a case study will follow, and finally I will assess . . .' and so on. The speaker has a wonderful opportunity at the start of a presentation. Most of the audience, generally speaking, will feel goodwill (being for the most part profoundly grateful that they are not doing the talking), and a good clear introduction will reinforce this feeling. The speaker has the audience's full attention; this will not easily happen again, and it would be a pity to waste the chance to make a good opening impression. Yet far too often speakers ramble into their topics, even apologizing for being there, and allowing the audience's concentration to slip before anything worthwhile has been said.

'Start on a strong note' is good advice. Just because the audience is listening so attentively, they will remember what is said, especially if the speaker captures their interest by useful words such as 'you' and 'benefit'. Their appetite is whetted for what is to come, and although the high point of concentration cannot be maintained indefinitely, it certainly lasts longer if they feel it is going to be worth their while to listen.

Main body Needless to say, the structure given in the introduction must be followed as the speaker moves into the main section of the presentation. The pattern is emphasized by summing up from time to time ('paragraphing') in order to reassure the listeners ('So I've looked at, now let's consider . . .'). Visual aids can be introduced as appropriate (see page 116) to support the speaker through the material, especially as the audience starts to lose its high level of attention.

Two-thirds of the way through any presentation is likely to be the low point. The audience's concentration was intense at the beginning, and has slowly declined, though at times it may have received a boost, for instance from a good visual aid or from a change of speaker. By the two-thirds point it is quite low, and the speaker must take action to remedy this. There are various possibilities: a striking example or a case study might be appropriate, one or two particularly attractive visual aids, some mild audience involvement (perhaps a question which the speaker can answer if nobody else does, or a challenge in terms of what they would do with a particular problem), or, of course, a touch of humour.

Humour can be very effective in a presentation; it can also be a disaster.

There are two basic criteria: the humour must arise naturally from the subject under discussion—it must never be artificially introduced in the way that a comedian might set up a joke—and it must not offend any member of the audience. This may be difficult, as much humour depends on specific aspects of human nature which are not immediately identifiable, and the speaker cannot be expected to know enough about the audience to be sure that the humour is not on dangerous ground. Humour must be planned, so that it can be rigorously checked for relevance and for audience reaction. Interestingly, if the humour does offend someone, a 'cold spot' will be created; other members of the audience who are sitting nearby will sense this, and the speaker may for the rest of the session be aware of a lack of warmth from one area of the audience without having any idea why it exists.

It follows that off-the-cuff humour is particularly hazardous. The comment may strike the speaker as remarkably witty and apposite, but it has not been carefully weighed, and it may cause lasting problems. This may seem particularly humourless advice, and we do not suggest that humour is a bad thing, but it needs to be handled with great care. The speaker must also be able to time the humour well—few things are as flat as a badly timed joke. Some people can handle humour brilliantly and use it to great effect in their presentations, but others cannot; it is always better to be on the safe side where humour is concerned.

During this main section of the presentation, the speaker must be alert to the response of the audience, quick to pick up any lack of understanding or confusion. Eye contact is essential (see page 100) so that the speaker reacts to the audience and the audience is reassured that the presentation is for them (it is a sign of inexperienced speakers that they talk to themselves, getting into a huddle as far away as possible and apparently pretending that the audience is not there). In a long presentation, it may be sensible to have a break for questions, although this must be carefully handled or control of the timing will be lost—there are few more serious problems for a speaker than to realize that time will run out before the final point is reached. Generally speaking, if the presentation is to last up to half an hour, it is better to take all the questions at the end; if it is to last much longer, then it may be helpful to have a short (maybe just five minute) break for immediate questions in the middle. Making a decision about this is part of the planning, and letting the audience know is part of the introduction.

The conclusion The main part of the presentation is completed, and the speaker moves towards the end. The audience should be told. Expressions such as 'finally', 'in conclusion', 'the last point' alert listeners to what is happening, and their concentration level at once rises steeply. The speaker must indeed be coming to a conclusion, and should then make the most of this last opportunity to impress the audience. The main points of the talk can be summarized briefly, and the most important emphasized: it is useful in preparation to imagine (perhaps not unrealistically) that the audience will remember only one point from the whole event; the speaker can then decide what that one point should be, and can highlight it at the end.

It is essential not to waffle to a standstill; too many speakers leave their audiences with 'well, that's it', or 'that's all we've got time for', neither of which is a very impressive conclusion. Make a strong and interesting final point, smile at the audience, thank them, and ask for questions. They may well remember the highlight of the talk long enough to fulfil their—and your—objectives.

At the end of the presentation, the speaker or the person in the chair will ask for questions; at the end of the question and answer session, the speaker has finished, but is still in front of the audience. The air of professionalism with which the presentation started must be maintained right to the end, or even afterwards, when members of the audience may waylay the speaker to ask for further clarification, to disagree or, with luck, to congratulate. The relief which most speakers feel at having finished the session must not be allowed to show until there are no members of the audience around and the speaker can finally leave the building.

Key points

- Identify the audience in terms of its expertise, experience, seniority and familiarity with the English language. Check the number of people who will be present
- Consider both the audience's objectives and your own
- Be professional, in appearance, confidence and commitment
- Know your own style and choose an appropriate level of formality
- Prepare the answers to questions, and the way in which they will be handled
- Plan an introduction which is informative about the speaker(s) and the structure of what follows
- Start on a strong note
- Guide the audience through the logical structure
- Encourage the audience's concentration, especially two-thirds of the way through
- Conclude on a strong note
- Be professional throughout the occasion

SPEAKING AT CONFERENCES

Conferences are occasions for a specialized form of presentation: they are sometimes organized by an area of industry or by a professional institution; they may be academic in nature, attracting both experts with world-wide reputations and young research students who want to make a mark in the hope of attracting research funding and/or university posts.

In each case, they are highly formal events with their own traditions; what has been said about making presentations applies, but there are particular constraints on speakers at conferences. Such occasions are showcases, often for both speakers and audience. They are often also opportunities to meet and to talk with colleagues and fellow workers from all over the world. This is valuable, but it means that many of the audience will not have English as a first language, although it is unlikely that they would attend if they did not have reasonable ability in understanding, if not in speaking it.

The language barrier

The prospective speaker must find out, usually from the conference organizers, if there is likely to be a language barrier. If a majority of the people attending will not be at home in English, or if the speaker has been invited to speak abroad, it is important to remember that the speed of delivery must be reduced and that body language may be suspect—there are parts of the world in which making eye contact, essential to a British audience, is considered impolite (see page 99). Visual aids will be particularly important, as will handout material—the more information given in written or diagrammatic form, the better.

Words must be chosen with care: all colloquial or slang expressions or non-technical analogies are better avoided, while technical language, which is more widely recognized by experts in the field, should be used as far as possible. If the talk is being translated, the speaker must pause to allow the words to be received; if simultaneous translation is used, a copy of the paper in English will be needed for the translator.

Preparation

A conference speaker will be given useful information by the organizers, such as the numbers expected, the length of the contribution (and how much of the time should be allowed for questions) and the equipment available. The venue is important, and speakers can use the checklist below to make sure that all the relevant details are covered. There is usually plenty of time to prepare the paper, as conferences are often organized a year or more before they take place.

Academic conferences

Many academic conferences expect speakers to read a paper, literally, and then to present the disk to the organizers so that the paper can be published. This seems to be both a waste of time—why read a paper in person when the audience could read it for themselves much more easily without the bother of going to the conference?—and inefficient, as the spoken language is very different from the written language, and most speakers would welcome the chance to revise their papers before committing themselves to print. There is obviously an advantage in meeting other researchers, but the opportunities given for discussing the work are under-used.

A more useful version of the academic conference is that in which the speakers send their papers out in advance, and their conference 'slot' is then used for a guided discussion, the speaker updating the information and the audience asking questions. This requires a much deeper knowledge on the part of the person writing the paper, as likely questions must be considered and answers prepared thoroughly in advance. Perhaps fortunately, it may be easier to foresee questions at this kind of conference, as delegates and their areas of interest are often known to the speaker.

Notes as suggested in Chapter 6 (page 106) are not always appropriate to the presentation of highly complex research material, and the speaker may decide that the text should be written out in full. If possible, complete sentences should be avoided as they are a temptation to reading, but fuller notes than usual can be written, with plenty of space on the page and the key points highlighted. A lectern may be provided: this will present a barrier between speaker and audience and may have the effect of rooting the speaker to the spot, but it cannot always be avoided, especially if it contains the only source of light by which the notes may be seen. Thorough rehearsal should still allow the speaker to make eye contact, but it will inevitably be more limited than at other kinds of presentation.

The speaker is more dependent on originality of material and the intellectual excitement it generates than on other occasions, and the talk should of course be tried out on colleagues for their comment and assessment. The timing is likely to be critical, as some conference programmes place three or even more speakers one after the other without a break.

Expert members of the audience may ask lengthy questions, call the basis of the speaker's work into question, or try to give a mini-presentation of their own. The person in the chair should be able to control this, but it is essential that the speaker remains calm and courteous, suggesting a private discussion later if it seems sensible to put a stop to the public debate.

New researchers may be invited first to give a short 'taster' rather than a full-length paper. Such a presentation, often no more than 10 minutes, is an excellent opportunity to get the feel of conference speaking and to gain confidence for the future. It is worth including one or two good quality visual aids, as they will make an impact which may stay in the audience's memory.

Perhaps the most difficult task of all is to assess the expertise of the audience. It is likely to be highly knowledgeable and sensitive to being patronized; speak in as much depth as possible but be prepared to give some additional help or explanation, perhaps as a handout, to those who lack detailed knowledge of the particular topic. It is acceptable to use the expertise available, for instance, if the speaker does not know the answer to a question, the audience can be asked for help. Most experts, thus appealed to, will be happy to offer their own ideas, and the speaker will appear to be open-minded, willing to learn, and suitably humble.

It is an honour to be asked to give a paper at a conference, and the speaker should take heart from the respect shown to his or her work; it can also be something of an ordeal, and it is wise to seek the advice of a more experienced speaker and to make sure that nothing in the preparation for the event is left to chance.

Venue checklist

Conference organizers should give the appropriate information to speakers well before the event, but they are not always as responsive to requests for information as they should be. The following checklist will help to ensure that speakers are not distracted unnecessarily at the point when they need to give all their concentration to the actual delivery of their presentations. It is divided into four sections, to show the stage at which each set of questions should be asked.

As early as possible
How many people will be present and how big is the room?
What visual aids equipment is available for me?
Can I dim the lights/use blackout?
Are there any visibility/audibility restrictions which I should know about?
Are adaptors, extension leads, etc. available if needed?
Can I check the equipment on the previous evening/before the morning session?

On arrival
Are the arrangements still the same as agreed?

Before the session
Can I use the equipment provided?
Am I at ease with the positioning of equipment, flexes, extension leads, etc.?
How do I find technical assistance if a problem arises?
Where should I stand?
Are the acoustics satisfactory/should I use a microphone?
Is there any visibility problem, for me or for the audience?

During the session
Are previous speakers having difficulty with equipment/position/acoustics?
If so, can I make things easier for myself?
Have any previous speakers created a potential problem for me, for instance by moving equipment?

A speaker who follows this checklist should feel confident that, whatever the difficulties of actually making the presentation may be, at least the venue will provide no unforeseen problems whatever!

Key points

- If the majority language is not English, speak slowly, avoid colloquial expressions, and use as much visual material as possible
- If the complexity of the material means that you want most of the text in front of you, use colour and space, and avoid writing in full sentences
- Rehearse until you can make eye contact with the audience as often as possible
- Ask colleagues to assess the originality and impact of your paper
- If you have no experience of conference speaking, take the opportunity to give a short 'taster' paper if possible
- Never underestimate the expertise of the audience; use it, if necessary

Conference venues

- Use the checklist provided (page 134), as early as possible in the preparations, when you arrive at the venue, just before the session and while other people are speaking
- Rehearse in the venue if possible, or at least try out the acoustics and the visual aids before your session

SPEAKING AT MEETINGS

Much has been written about meetings, especially the psychological aspects, and we are not here concerned with the games people play in order to influence decisions; this section will concentrate on effective speaking at meetings, from the point of view of the person who takes the chair (hereafter called the chairperson) and the technical specialist who has to spend a good deal of time—often reluctantly—attending meetings.

For some people, it is more stressful to speak at a meeting than to give a formal presentation. It is harder to plan a contribution, as others will also have their points of view to put forward, and people are generally in close proximity to one another. The extra formality, indeed the status, given by a platform is missing, and people have on the whole to fend for themselves.

Speaking from the chair

The chairperson has an important role to play in guiding individual contributions and in shaping the meeting as a whole. Preparation is essential. In thinking of the venue, it must be made clear that the meeting may be interrupted only in the most dire emergency: messages or telephone calls interrupt the flow of proceedings and disconcert a speaker. In planning the agenda, the chairperson must consider not only a logical and helpful order of topics, but also who is likely to want to speak on each issue. This means knowing the members of the meeting well enough to be aware of their interests and to ensure that everybody who has a vested interest in each subject has a reasonable chance to speak. It also means knowing who is likely to be controversial or long-winded, and where factional interests might show themselves. Perhaps two of the most important considerations at a meeting are that everyone who wants to speak has a chance to do so, and that the meeting does not go on for too long; it is difficult to get the balance right.

The agenda will have been drawn up well in advance of the meeting (see meeting documentation, page 71); the chairperson and the minute writer will have discussed the timing in some detail, identifying the major items for discussion and those which can be dealt with briefly. At the beginning of the meeting, the chairperson will let everyone know what has been decided and how the meeting is planned. New members of the group will be introduced and welcomed, and if necessary each person attending should introduce himself or herself. It is essential that everyone knows who everyone else is; if people come from different organizations, it is a good idea to give out nameplates to be set out on the table so that nobody has the advantage over others of being able to address colleagues by name. The chairperson may also describe what the meeting hopes to achieve, and briefly introduce the first item on the agenda.

Once the meeting has started, the chairperson is in control (tactfully) and responsible for making sure that the plan is adhered to. People must be encouraged to speak, by name if necessary, and to follow the traditional formality of meetings by addressing the chair rather than each other. This has the added benefit of 'distancing', of making emotional input seem less personal; the chairperson must ensure that no issue gives rise to abuse or excessive antagonism. Body language (see page 99) often provides a clue sooner than words: if a member is drumming fingers on the table, tapping feet, using hands in an aggressive way, scowling and so on, some cooling off is needed (perhaps by the perpetrator being given a chance to air the grievance); on the other hand, leaning back in the chair, folding arms or staring out of the window quickly registers boredom, a 'switching off' from the activity of the group. Addressing a question to this uninterested member by name will usually bring rapidly-renewed attention.

The chairperson must take care not to speak too much or too quickly. It can be difficult to restrain a strongly held point of view or violent disagreement with what has been said, but a breach of the traditional impartiality of the person in the chair may cause yet more ill-feeling. If necessary, the chairperson may give an opinion, but tactfully, making it clear that this is just an opinion and allowing plenty of time for contrary views to be discussed. Time may be needed for finding the right page or for making notes, and the chairperson should pause and check that everyone is ready before moving on. Eye contact and a readiness to smile are important in chairing meetings as in every other kind of presentation.

Visual aids are unusual in meetings, but a flipchart can be invaluable to a small group. Ideas or problems to be solved can be written up in just a few key words so that everyone can concentrate on them; a course of action may be drawn as a flowchart so that people can see how their own contributions fit into the whole; the pros and cons of a situation may be listed for consideration. Having something to look at helps those attending to focus on the subject under discussion rather than allowing their minds to wander.

From time to time, the chairperson should briefly sum up what has been decided so far; this encourages everybody by showing what has been achieved, and clarifies the details in people's minds. It is also a good idea to check the wording of any agreement (including the name of any person charged with taking action) so that the secretary or other minute taker is in no doubt, either. If there is a query, it is much easier to deal with it while all the members of the meeting are present than to allow it to re-emerge under 'matters arising' at the next meeting. The chairperson carries great responsibility for making sure that what is recorded is an accurate reflection of what took place, and at the same time for reassuring those present that their views have been adequately expressed. (See also minute writing, page 73.)

The meeting should end within five minutes of the agreed time, and the chairperson can close with thanks and, if possible, a compliment to everyone present. 'We've really made progress today, splendid' will encourage people to feel positive about the meeting and maybe to come to the next one.

One more duty of the chairperson may be to prepare the seating plan for the meeting. As far as possible, everyone present should be able to see everyone else. Eye contact is as important at a meeting as it is in a presentation, and people will find it easier to contribute if they can see and be seen. It is also possible to react quickly to diffuse aggression or ill temper if the first signals are picked up quickly; those who are known to hold opposing views may co-operate more easily if they are side by side rather than sitting opposite one another in a position suggesting confrontation. The chairperson's main object is to encourage everyone to speak in a positive, courteous way, and this is more easily achieved if the eye contact is good.

Reporting to the meeting

Technical people are asked from time to time to introduce a subject for discussion, perhaps reporting on progress so far or giving a short report which is the basis for an item on the agenda. Reading half a dozen pages, full of technical detail, is not advisable. It is much better to circulate the paper in advance, so that everyone can absorb the information, and then to repeat the highlights to the meeting, stressing decisions which have to be made, money that has to be allocated, or similar actions which are the prerogative of the meeting.

Even so, the speaker may be addressing the meeting for several minutes, and decisions have to be made. Should the speaker sit or stand? Generally, if the person speaking wants to make eye contact with everyone else, it is necessary to stand; this is also helpful if visual aids are to be used. If the group is very small and the speaker would appear to dominate, it might be better to sit. Notes (see page 106) are better held in the hand than put on the table, and the speaker should use the same body language as in any other presentation, though modified if the audience is small and physically much closer.

Such presentations will of necessity be short—other people want to join in the discussion—but they should be prepared and rehearsed in the appropriate way. Questions are sure to follow, and the wise speaker will have prepared the answers, having facts and figures available or, at the worst, being able to find the details for the next meeting.

Speaking at the meeting

Most professional people spend what most of them would consider too much time in meetings: formal meetings with other companies are stressful as well as unavoidable; meetings, formal or informal, within one's own company are time-consuming and, very often, more or less impromptu, so that the individual goes along with something to say and a lot of time in which to listen.

Positive listening is helpful to the progress of the meeting. If those attending look interested, take notes and ask questions, it is unlikely that the meeting will be seen as a waste of time; if nobody appears to be involved with the topic under discussion, then the meeting flounders and may ramble on for far too long. However much the individual resents having to attend, a positive attitude is helpful to everyone else. There may well come a moment when the disgruntled arrival has something to say.

A contribution can almost certainly be prepared in advance. It is difficult to guess exactly how the topic will be presented, but it is reasonable to work out a point of view and the main thrust of the argument, and to decide on the action required as a result of the intervention. The use of notes will give the contribution weight. Careful listening will also reveal who will be in agreement with the points raised and who will be against; it is a good technique to include references to other people's words in order to encourage support. 'As Jim said earlier', 'If I can go on from the point made a few minutes ago by Sarah', 'I really support what Philip said, and I'd like to add ...' are useful ways of building on what is already agreed.

There is a great temptation to rush; perhaps people feel, rightly, that the time allocated will be brief, but speaking slowly, pausing for thought and obviously weighing every word, creates an impressive effect—this contribution comes not on the spur of the moment but from careful thought. If the meeting is formal, all interventions should be through the chair, and speakers must absolutely refuse to respond to aggression. A moment's thought before replying to questions is helpful, as is eye contact with the questioner (although if the question seems to be aggressive, there is much to be said for avoiding eye contact—it is surprisingly difficult to attack someone who is looking elsewhere). We have all been aware of people who are listened to at meetings, who do not often contribute but whose words are always taken seriously; they take their time—and never waste it.

Key points

Guidelines for the chairperson

- Make sure that the meeting will not be interrupted
- Find out as much as possible about those who will be attending the meeting
- Introduce new members, or ask them to introduce themselves; use nameplates if possible
- Encourage everyone to participate, and watch for revealing body language
- Do not intervene too often, get carried away by your own opinions, or speak too quickly
- Use a flipchart as an aid to discussion
- Close the meeting decisively, thanking everyone for coming

Guidelines for people at meetings

- If you have a lot to say, circulate a paper in advance and speak to it briefly at the meeting
- Be ready to answer questions
- Listen positively, making notes and asking questions as appropriate
- Build on previous contributions, especially where there is agreement
- Take your time, but keep your intervention brief and to the point

VIDEOCONFERENCING

Meetings may be particularly wasteful in time and money if the company operates on widely separated sites. Key personnel can be faced with four or five hours' drive for a meeting lasting hardly more than an hour, resulting in stress and frustration in addition to the unproductive time spent on the road. Videoconferencing provides a partial answer; it is extremely expensive to set up, but in the long term may save money and free staff for other work.

A dedicated room will be needed at each site, as the video monitors must be large scale to allow for up to six participants at each end, and an overhead projector with camera above it is useful for the transmission of documents. Each group sits at one side of a large table, facing the screen, and twin cameras give a long panoramic view, although the remote control allows close-up of an individual speaker. The room itself should be as soundproof as possible, with simple, uncluttered furnishings (plain-coloured full-length curtains are best) and adequate artificial lighting, easier to control and more even than daylight.

Videoconferencing is ideal for short meetings; it is less satisfactory for meetings which are likely to last for more than a couple of hours, when the restrictions of movement and view can become irksome. As the time has to be booked in advance, other staff may be annoyed at not having access to the room for long periods—thus losing some of the benefits of the system. Once staff are familiar with the style needed and lose the inevitable self-consciousness of the first-time participant, they will see the benefits of regular short discussions with colleagues perhaps hundreds of miles away, without anyone having to leave the building.

Some guidance should be given when the system is first used. A videoconference has to be more tightly structured and controlled than other meetings, often beginning with a formal statement from each side, not least to allow the participants to relax. No matter how heated the discussion, people must speak one at a time, or the sound will be unclear at the other end. Inevitably, this results in some loss of spontaneity; asides, interruptions or subordinate conversations will not be heard but may make the main speaker inaudible. It is also harder to judge reactions: if the camera has closed up on one speaker, signals of disagreement or boredom among other participants will not be seen. In any case, as the view is restricted to the table and above, foot-tapping, for instance, would be hidden.

Although diagrams can be shown through the overhead projector, and the resolution is good, the meeting is slowed down while everyone reads the screen; the discussion of detailed technical material is difficult, and it is usually better to fax a diagram first, so that each participant has a photocopy on the table for reference.

In spite of these limitations, companies which use videoconferencing find that it becomes very popular as a relatively painless way of holding meetings. Indeed, it has another hidden benefit—because the time has to be booked and is expensive, discussions tend to be shorter and better focused than they are when those involved are recovering from a long journey or greeting people they have not seen for some time. There is much less temptation to ramble!

However, a word of warning is needed. There is little security in a videoconference; it is too easy to eavesdrop electronically. Videoconferencing, for all its benefits, should not be used for the transmission of confidential or sensitive material.

Key points

- Videoconferencing saves time and stress, and for many companies is well worth the cost
- You need a dedicated room which is sound-proofed, simply furnished and with good artificial light
- Use videoconferencing for short meetings, and allow a few minutes for the participants to relax
- Good control of the meeting is needed, so that people speak in turn
- Some body language is lost—you need to be particularly aware of the restrictions on movement, and of distractions
- Fax detailed information or diagrams first, so that everyone has a copy
- Videoconferencing is expensive, and a fair system of booking is essential
- Do not use videoconferencing for confidential discussions—it is not safe!

SPEAKING ON THE TELEPHONE

Most people in the United Kingdom have a telephone at home, and it might be expected that everybody could use it efficiently at work. Most of us know that this is far from the truth. Practical difficulties such as being unable to reach a pen, paper or diary can cause an awkward hesitation, although this is as nothing in comparison to being put through to the wrong person, perhaps several times, or leaving a message and not being rung back. It cannot be overstressed that the telephone is still the front line at work, and that a poor or discourteous reception can lose clients. This is perhaps one argument against the use of the car phone, even when the vehicle is stationary: the speaker is removed from a working environment and is less likely to consider the right approach to the conversation. This may also be true of mobile phones used at other times, for instance during a train journey.

Professional staff should be aware of all these problems, and if they are not available should leave a clear message about where they are to be found, how long they will be out of the office and when they can ring back. Most telephone extensions can be put through to a different number automatically so that a message can be left, but the person responding should be able to give adequate information about the intended recipient of the call. If there is a delay, the hopeful speaker is often presented with music, which will either soothe or drive the listener mad, depending on temperament. At least it shows that the line is still connected.

Telephone calls are by their nature ephemeral, and if they result in useful information, it is important to make notes. Telephone pads and pens tend to disappear or, as we said earlier, to be out of reach, but ideally there should always be a means of writing nearby, so that a written record may be made. This is true whether the person who takes the call is the intended recipient or someone else taking a message. Human memories are notoriously fallible.

Fax machines (see page 38) have made telephone calls easier. It is possible to send technical details or diagrams by fax or e-mail, and then to ring to discuss the response or to ask questions. We can also confirm or elaborate on the call, with little or no delay. In some ways, the possibilities of the telephone have been increased, as there is no need in sending a fax to work out what the time is at the other end, and the message can include a sensible time at which a telephone discussion can take place.

Telephone calls are expensive, and a few minutes' planning beforehand will save time and money. At the same time, they are personal, and a moment or two spent in greeting, asking about the well-being of the recipient or even commenting on the weather will not be wasted if it produces increased goodwill. Needless to say, the caller needs to be quickly responsive to any sense of urgency or hurry at the other end of the line—there may be no visible body language, but the tone of voice can give as clear a message as the words used!

There are two varieties of call, those we initiate and those to which we respond. The former clearly put the speaker in a strong position, provided that he or she has prepared for the call. The name, job title and company name of the recipient should be in front of the speaker, not just the necessary telephone number; it may also be a pleasant courtesy to have the name of the personal assistant who is likely to answer. If the call is a complex one, perhaps including technical data, notes are needed so that no detail is forgotten. Numbers, prices and the like should always be in writing, so that there is no chance of error. Since it is difficult to be sure of such details over the telephone, it may be worth following up the call with faxed information.

Responding to someone else's call is much more difficult. There is no opportunity to collect either thoughts or material, and only the briefest of moments in which to adopt the right posture. It would be foolish under these circumstances to commit oneself to giving exact figures or technical data, and it is almost always better to respond brightly and promise to ring back with the required information in half an hour. If the question can be answered after a moment's thought, then allow the moment, although as the caller cannot see the thinking process, it is as well to say, 'Just a moment, please, I'd like to think about that.' A short pause is acceptable, but it must not be prolonged until the caller wonders whether there is still anyone there.

Body language and the telephone

Some calls are of particular importance to the speaker: those involving clients, potential customers, prospective employers and so on should be very carefully prepared, and the caller should try to get into a positive, efficient frame of mind before ringing. Posture is surprisingly important. We may be unaware of the fact, but we respond very differently to telephone calls: if we are at home, for instance, and receive a call from a friend, we probably slouch on the chair, pick up something from the telephone table to fiddle with, and generally make ourselves comfortable (we may even be drinking a cup of tea at the time). If the call is from a senior manager at work, we instinctively sit upright and assume a businesslike expression and attitude, even though the person at the other end cannot see us. In the same way, it is a good idea to prepare for an important call by sitting in a comfortable but upright posture, so that we feel alert and can breath well (the importance of good breathing is dealt with elsewhere, page 104).

Smiling, as we have also noted (page 98), affects the voice. If we smile at the person at the other end of the line as we introduce ourselves, the voice is 'brightened' and sounds more encouraging and easy to listen to. A complex telephone call can be almost a small presentation, and we should think of it in that way. It is certainly a means of creating goodwill for ourselves and our company.

The answerphone

Unfortunately for many people, a call may bring not the expected response but an answerphone. These useful machines are often dreaded and certainly often misused. It is not worth trying to leave a lengthy or complex message, but some information should always be given: the name of the speaker, the company and the telephone number, and a brief indication of what the call is about. Two particular items should be given slowly and clearly—the name and the number. Far too many people rattle off their own telephone number at high speed, forgetting that the listener has not only to hear it but to use it; having to replay the message three or four times to try to catch a muttered or rushed number is infuriating (and a frequent irritation). Speak fairly slowly to the machine, give numbers clearly (in English, this usually means in groups of not more than three digits as far as possible) and a call back will almost certainly result.

Key points

- When using the telephone, make sure that pen, paper and diary are always within reach; beware of involved conversations on mobile or car phones
- Leave adequate information about your own movements, and when you will be available
- Make notes, especially of technical data, and follow up the call with a fax or letter of confirmation to reassure the recipient

- Prepare your attitude and your posture to give a businesslike response
- Smile at the listener, to brighten your voice
- When you receive a call, it is better to ring back later than to give inaccurate or inadequate information
- Be appropriately friendly, courteous and responsive to the person at the other end of the line
- Speak slowly and clearly to an answerphone
- A written record is much more likely to be accurate than human memory

8

Specimen Report and Presentation

Background • Preparing the report • Presenting at the seminar • Presenting to clients

In this chapter, we follow a particular engineer through the preparation of a report on work carried out under contract for a commercial organization; we see how he prepares for a seminar in his university department, and subsequently for a formal presentation to his sponsors. The material is based on a real-life project carried out some years ago. Sections of the report are given as examples, as are some of the diagrams; the process of preparing both presentations is discussed in detail.

BACKGROUND

Moroges et Jambles (known to its personnel in the UK as M & J) is a French glass manufacturer with subsidiaries in various European countries including the UK. The British company has close links with nearby Abimouth University, and from time to time sponsors research projects of interest to its work.

The company has become concerned about the effects of weathering on the strength of window glass; the subject has been drawn to its attention because of the increased stress on glass caused by heavy traffic and sometimes by quarrying, road building and similar operations which cause vibration of windows. Andrew Poynter, Senior Lecturer in Mechanical Engineering, who has acted as a consultant for M & J on several previous occasions, has been asked to look into the problem.

Andrew's brief is to carry out an investigation into changes over time in the strength of domestic window glass because of weathering, in situations where it will be subjected to vibration. Initially, the work will concentrate on the short term, that is, possible deterioration within the first 12 months.

Andrew is asked to prepare a report, and then to explain the results in a short presentation to M & J staff.

PREPARING THE REPORT

Andrew identifies his terms of reference as follows:

TERMS OF REFERENCE

The strength of glass in domestic windows has become a subject of concern because of the increased vibration caused by extensive road works. This problem could be exacerbated by any loss of strength in the glass as a result of natural weathering.

Andrew Poynter, Senior Lecturer in Mechanical Engineering at Abimouth University, has therefore been asked to carry out a preliminary investigation to discover the loss of strength, if any, caused by the weathering of the glass in normal use.

Andrew accepts the terms of the agreement. Before beginning, he clarifies the objectives of the report he will eventually write, using his discussions with M & J staff to identify the readership and their likely areas of interest. This information will be important to him as he plans and writes up the results of his research, and so he notes it for future reference, as follows:

READERSHIP AND OBJECTIVES

The report will be read by interested M & J staff with technical expertise in the field; it may also be of interest to more senior staff, for instance with financial responsibilities; the report may also go to staff in the parent company, who will probably be able to read it in English but for whom English is not the first language.

The objectives of the research are:

1 to investigate the possibility of a loss of strength in 3 mm domestic window glass as a result of weathering
2 to quantify any such loss in the short term, that is, within the first year of use.

Having identified his readership and objectives, Andrew can put the report out of his mind and concentrate on carrying out the research—the aspect of the project which appeals to him most. Nevertheless, he drafts the basis of a spider diagram, to clarify the stages of the work and also to help him later when he has to organize his material. He is at the same time making decisions about the tests he will conduct, the equipment needed and the time to be allocated to this project. This spider base is illustrated in Fig. 8.1.

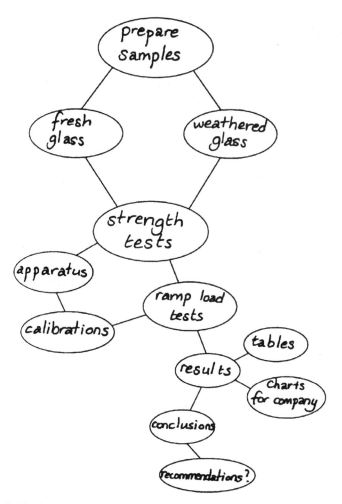

Figure 8.1 Spider base

As the final stage of preparation, Andrew uses the spider, together with his own knowledge of similar documents in the past, to draft a Contents List. It follows a common pattern for laboratory test reports.

1	Introduction
2	Apparatus
3	Method
4	Results
5	Conclusions
6	Recommendations

A summary will of course be prepared as the last stage of the report writing, and placed at the beginning of the document. At this stage, Andrew is unsure about the need for a Recommendations section, and decides to consider this later.

Andrew now carries out his tests, adding a few brief notes to his spider diagram from time to time in case he forgets the details. The results are recorded and he analyses them, drawing the appropriate conclusions. His computer screen is now showing an impressive array of information, and, with a sigh, Andrew realizes that he must start to organize it all into a report and subsequently into a presentation for M & J staff. He goes back yet again to the spider, adding extra notes at the appropriate points. It now looks like Fig. 8.2.

Figure 8.2 Full spider

The first section in the main body of the report will of course be the Introduction. Andrew hesitates about this, as in previous reports he has found it a difficult section and has frequently put off writing it until near the end. However, he decides that in this case he can draft it fairly easily, and so he does so.

---

1 INTRODUCTION

The increased risk of vibration damage to domestic window glass has called into question the extent to which the strength of such glass is affected by the natural process of weathering. A preliminary investigation was therefore commissioned by Moroges et Jambles from Andrew Poynter, Senior Lecturer in Mechanical Engineering at the University of Abimouth. The research was limited to 3 mm domestic glass and the extent to which it was affected by weathering in the first year of use.

Tests have therefore been carried out on fresh batches of glass under conditions of natural weathering. The glass was supplied by the company and cut into the form of circular discs 15 cm in diameter.

As he looks back at his notes, Andrew sees that he had jotted down his thoughts about the three main effects of weathering: the surface of the glass reacts chemically with water, to produce caustic soda which damages the surface; abrasion is caused by windborne particles; devitrification causes the glass to crystallize and shrink—a process which is speeded up by impurities in the glass. This third effect is not likely to apply to the modern glass which Andrew has been investigating, and in any case it would not present a problem in the short term. Andrew feels that all this general information might not be needed in a report which would probably be seen mainly by specialists; he may perhaps mention the chemical reaction and the effect of abrasion when he finalizes the introduction, but devitrification is irrelevant to the glass he has tested.

Andrew looks again at the notes he made about his apparatus. He has carried out ramp load tests, as they seemed the most appropriate to normal window conditions, and he used a rig which had originally been built for another research project. He will need to include a schematic diagram of the hydraulic circuit, but otherwise the Apparatus section seems to be short and easy to prepare:

---

2 APPARATUS

A rig for the testing of circular glass discs under uniform pressure has previously been developed and built (for a schematic diagram, see Figure 2.1), and was used again in the tests described in this report. The stresses produced by this type of rig are virtually zero at the edges, so that edge breaks are avoided. Fractures start near the centre and are therefore a function of the surface condition of the glass.

A pressure transducer was mounted in the hydraulic line to the test vessel, connected to a computer which recorded and simultaneously displayed the results. The pressure rise time and the failure pressure were thus noted.

An aluminium disc was used as a calibrator for the tests. This was possible because aluminium has a similar Young's Modulus and Poisson's Ratio to glass, and was necessary because attaching a strain gauge to the glass would weaken it. From the experiment with the aluminium disc, look-up tables were created in the computer which enable calculation of strains to be made for any given pressure.

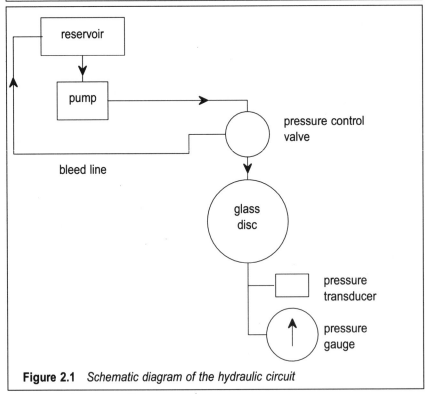

Figure 2.1   *Schematic diagram of the hydraulic circuit*

Figure 8.3   Apparatus diagram

The next section of the report which Andrew has to tackle is Method; again, the spider diagram shows him what to include and his notes fill in the details.

3 METHOD

The choice of test to measure the strength of the glass was determined by the need to simulate normal window load conditions as closely as possible. Ramp load tests fulfilled this requirement, and were comparatively simple to carry out. The principle of such tests is that the pressure can be uniformly increased at a predetermined rate until fracture occurs.

The tests were carried out in a room with a constant temperature of 20°C.

Two batches of glass were tested, as described below, and the results recorded.

3.1 Fresh glass (the control)

Fresh glass was obtained directly from the manufacturer and delivered in protective packaging. It was in the form of sheets, 1 m square and 3 mm thick.

Great care was taken to prevent surface damage during the preparation of the glass specimens. Constant 100% humidity was maintained by the use of wet cloths over the glass, and the cutting was performed on a bench protected by a blanket and by a double layer of polythene sheeting. This ensured that the surface of the bench was free from small slivers of glass, grit or dust.

The sheets of glass were cut into lengths of 1 m by 0.2 m, and then into 0.2 m squares. These squares were then cut into circles. Between each cutting operation, the glass was washed and stacked to dry on wooden racks; the table was vacuum cleaned, brushed and washed.

The glass circles were then tested on the rig, the pressure rise and failure pressure being recorded by the computer.

3.2 Weathered glass

Natural weathering of glass was carried out in order that a comparison might be made with the strength of the fresh glass.

Ready-cut discs of glass were placed on the roof of the University's Engineering Block, in wooden racks. The discs were marked with the date and left for a year before being taken to the laboratory to be tested on the rig. The racks were made of tongued and grooved board, sufficiently high off the ground to eliminate the chance of grit being splashed up during rainfall.

These discs were tested on the rig in the same way as the fresh glass discs.

Andrew has to present the detailed results of these tests and his statistical analysis in Section 4 of his report. He decides that the written explanation would be clarified by two tables, one for the new glass and the other for the weathered glass. These will be identified as Tables 4.1 and 4.2 respectively; they are given below, together with a graph to show the changes over time in the mean strength of the glass after weathering (see Fig. 8.4).

Table 4.1   *Results for new glass*

Sample	Number of specimens	Mean strength N/mm^2	Standard deviation N/mm^2
1	29	164.6	42.9
2	30	143.3	38.0
Total	59	153.8	40.2

Table 4.2   *Results for naturally weathered glass*

Number of days exposure	Number of specimens	Mean strength N/mm^2	Standard deviation N/mm^2	Significance of the differences between the batch and fresh glass
8	19	130.0	40.8	VHS
20	25	146.1	46.9	NS
34	24	127.8	32.4	VHS
90	22	129.0	15.0	HS
180	26	98.0	23.6	VHS
270	20	119.8	18.0	VHS
365	31	101.5	22.8	VHS

Note: Significance analysis
Comparisons with fresh glass are given above. The ordinary z-test has been made for the significance of the differences of means where the sample size is greater than 25. Smaller samples have been subjected to the t-test. The levels of significance are graded as follows:
NS – no significant difference
HS – highly significant difference
VHS – very highly significant difference

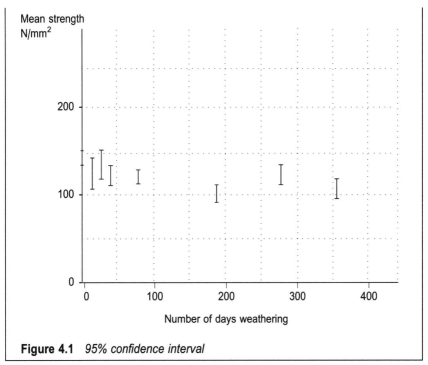

Figure 4.1 *95% confidence interval*

Figure 8.4 Results diagram

Andrew now considers his findings, and decides that they are interesting but in the long term inconclusive. Further tests are clearly needed—and he would like to have the opportunity to carry them out. He therefore decides that he will have the Recommendations section which he had considered earlier.

The two remaining sections are drafted as follows.

---

5 CONCLUSIONS

The tests described in this report have shown a progressive reduction in the strength of window glass over a period of one year, from a value of $150N/mm^2$ for new glass to about $110N/mm^2$ after one year's weathering.

On the basis of these tests, there is no evidence to show whether the process has levelled out.

6 RECOMMENDATIONS

The research carried out has shown some reduction of strength in the glass over a year. If such a reduction were to continue, there would be cause for concern.

It is therefore strongly recommended that more tests are carried out on data collected over a longer period of time.

Andrew now has most of his report more or less complete. He is, however, a little worried in case he has not given sufficient weight to the theoretical basis for the work, and he returns to his notes. Perhaps he should add a paragraph at the start of the Method section, to justify his treatment of the results: he has used a conventional formula in order to convert the uniaxial test results into the ultimate stress values which he has displayed, and he feels that he should show how this was done, in order to reassure readers who might not be familiar with the technique.

He therefore adds an appropriate paragraph immediately under the Method heading. There is also one reference to be added: the details of a book to which Andrew has referred must be given at the end of the report, and the appropriate mark made in the text.

Andrew now finalizes his Introduction, which is a far easier task now that much of the report is completed; he decides to include a brief comment about the two relevant causes of weathering, the chemical reaction and abrasion by windborne particles. Otherwise, the Introduction will remain much as it was first drafted. The Contents List has also to be revised, but this task will consist merely of adding the subheadings which have been used in the text.

Andrew now has to tackle what is in many ways the most difficult section of his report—the Summary. He therefore goes back to his Introduction, and his Conclusions and Recommendations, and links them with significant evidence; on this basis he produces the first version of the Summary. He deliberately makes this too long, knowing that it will be easier to cut it than to try to extend it later. His first draft reads as follows:

An increased risk of vibration damage to glass gave rise to a request from Moroges et Jambles for a preliminary investigation, to be carried out at the University of Abimouth, into the effect of weathering on 3 mm domestic glass. Ramp load tests, which simulate normal window load conditions closely, were carried out on fresh glass and repeated on weathered glass. The results of these tests showed a progressive reduction in the strength of the glass over one year's weathering from $150N/mm^2$ for new glass to about $110N/mm^2$ for the weathered glass. This reduction in strength, if it were to continue, would give cause for concern, and there was no indication that this process had levelled out. It is therefore recommended that further tests should be carried out, using data which reflect a longer period of weathering.

This includes Andrew's main points, but it seems wordy in style—the first sentence in particular seems to ramble. On reflection, he also feels that quoting the exact figures in the summary is inappropriate—he should simply be giving the principle of what he had found. He therefore revises the summary as follows.

SUMMARY

In response to a request from Moroges et Jambles, a preliminary investigation was carried out at the University of Abimouth into the effect of weathering on the strength of 3 mm domestic glass. Tests were carried out on both fresh and weathered glass. The results showed a reduction in the strength of the glass over one year's weathering; on the basis of these tests, there was no indication whether the process had levelled out.

Further tests are therefore recommended, using data collected over a longer period of time.

Andrew has now completed the writing of his report, but there is the process of checking and revision to be carried out. He uses his computer spell-check, reads through a printout, making various amendments, and asks a colleague to go through his draft. Only after this has been done can he finalize the text, add a title page and print out a good quality copy which can be bound and sent to the company that sponsored the work.

PRESENTING AT THE SEMINAR

The report is sent off to M & J, and some time later Andrew is invited to the company's UK headquarters in order to give a short presentation about his research. He is aware of the importance of this occasion in the light of his recommendation that further work should be carried out, and he looks at once for an opportunity to present his material to colleagues and to ask for useful comments.

The Engineering Department holds an informal lunchtime seminar each month, partly to give staff and research students the chance to meet and to discuss their work, and partly to give new researchers experience in speaking to a group. Andrew asks for and is readily given a 10-minute slot at the next seminar, in which to make a preliminary presentation. As this is in a few days' time, he knows that he will not have his formal visual aids ready, but he reckons that the overhead projector will do; on second thoughts, he decides that a demonstration might also be feasible, and arranges with a technician to have the benchtop rig brought in from the laboratory. He is aware of the dangers of carrying out a demonstration in front of an audience, but in this case he knows everyone well and feels that they will be sympathetic if something goes wrong—and that he can take the inevitable teasing.

Andrew considers what he wants to get out of this occasion, and decides that he wants technical criticism, some idea of the questions he might be asked in-company, and a rough idea of how much he can reasonably say in the time available. The audience is coming because of their general interest in the subject; they are likely to enjoy seeing a demonstration of the research, and as far as the results are concerned, they will want to know the trend but probably not the detailed statistics.

In the light of this last consideration, Andrew feels that he should show a version of the graph which he used in his report (see page 153). When he looks at it again, he sees that the print is too small to be copied and projected satisfactorily, there are irrelevancies such as the figure number on the page, and the grid lines would look untidy and be difficult to reproduce accurately; it cannot be printed directly onto acetate. Fortunately, it is still on Andrew's computer and, with a little tidying up, it can be transformed into a suitable visual aid.

Using his report and also his spider diagrams, Andrew starts to prepare his presentation. There are usually a dozen or so people attending the seminars, and so there should be no problem in their seeing the rig in operation. Andrew decides to take with him the aluminium disc with the strain gauges on it which he had used for calibration, and a couple of glass discs, one of fresh and one of weathered glass, to use as his specimens. As he is thinking about this, he has a bright idea. He could back a disc with sticky brown paper, and then put it through the rig. When it fractures, the pieces will not fall apart, and the pattern of breakage will be there for people to see. If this works well at the seminar, he might take it to the company presentation as well.

Andrew has only 10 minutes, and he has therefore to be very selective in what he says. He will have to explain how he came to be doing the research and his links with the company; that will form a very brief introduction, but everybody at the seminar will know him, and will be aware of his specialist research. He will need to explain about the calibration, and how he aged the glass, but most of the time will be taken up by the demonstration itself. He realizes that this will not be possible in his major presentation, and plans to have slides made of the rig itself; the seminar is already proving its usefulness by drawing his attention to what the audience needs to see and the most appropriate way of presenting the information to different audiences.

The seminar group will need to be told about the statistical analysis (but very briefly, with the graph on the screen), and of course about his conclusions; this part of the subject can be rounded off fairly quickly, and the possibility of further research will go down well with his colleagues. The theoretical part of the work will have to be left out, but Andrew wonders if he might get questions about the equivalent stress. He therefore thinks in advance about how he can give a clear, brief answer.

It is difficult to rehearse without the equipment in place, but in the laboratory Andrew tries out what he wants to say, and goes through the demonstration stage by stage without actually using the rig. Inevitably, he finds that he is going to overrun, but it is not difficult to cut back on his description of the mechanical effects of weathering—he is again likely to be able to fill in the details in answer to a question. He soon estimates that he has his presentation down to the right length, knowing that the timing will in any case have to be much more exact when he visits the company than it needs to be when he talks to such an informal group. He also decides that on this occasion he will not use notes, relying on the equipment and the one overhead transparency as prompts. As he makes this decision, he thinks ruefully that he will feel very differently when he goes to the company.

The seminar duly takes place; Andrew's demonstration works well, and there is considerable interest on the part of the audience. There are a number of questions, including the one which he had foreseen about equivalent uniaxial stress. He is greatly relieved that he had anticipated this, and feels that he can give an appropriate answer. A more difficult question, however, is: 'What are the major mechanical effects of weathering, and how do they affect strength?' This is a wide question, and Andrew has little time in which to answer it; he makes a quick decision to concentrate on abrasion as the start of the process of crack development, stressing that this is simply an example. His answer seems to satisfy the questioner.

Later that day, Andrew reviews what he has learnt from the seminar. He realizes yet again how little can be said in 10 minutes, or even in the 20 minutes which the company is giving him. He will again have to be selective in organizing the material, and as timing will be critical, he will need notes and will rehearse the presentation several times. He can also now clarify what the audience will want to see; he will not be able to take the rig, and so he must have some good quality slides to show the equipment—perhaps not just the rig, but also the glass in racks on the roof during the weathering process. It looks as if he must also give more thought to explaining the mechanical effects of weathering, but this may depend on his audience. He must find out more about them.

All in all, the seminar has proved to be a valuable learning experience; Andrew has also enjoyed it, and the audience has appreciated the chance to see his experiment in action.

PRESENTING TO CLIENTS

Term is ending, and Andrew at last has time to prepare for his presentation to M & J. He has been told very little apart from the time available—20 minutes for the talk, followed by 10 minutes for questions—and the venue, which is the Small Conference Room at the company's UK Head Office. He must find out more, and so he rings the Personal Assistant to the Technical Director, whom he knows from their earlier meetings when the research project started. He leaves a series of questions, and, within two days, he receives a fax with the answers.

There will, it seems, be about 10 people in the audience. As well as the Technical Director and two or three other technical staff, one of whom he also knows, there will be the Contracts Director, whom he has met, someone from the legal department, and, to Andrew's surprise, the Managing Director, about whom he knows nothing except that he originally trained as an accountant. There is another unforeseen development: two senior technical staff from the French parent company will be visiting the site and have been invited to come to the presentation. One of them speaks excellent English, but the other has only limited knowledge of the language, although he can follow technical material reasonably well.

The room itself holds perhaps 50 or 60 people. It is hexagonal in shape, air-conditioned, with blinds and lights which can be dimmed. Good modern visual aid equipment is built in; the audience will be seated in a semicircle with a clear view of both speaker and equipment.

Andrew feels that he will have all the facilities he needs to give a good presentation, and in the light of the audience, he knows that everything must be of the highest quality. He visits the University's Audio-Visual Unit to get expert advice about his visual aids, and explains the need to use slides to replace the impact of his demonstration. After some discussion, they all agree that he needs a slide of the equipment itself, but, as the glass discs are small, that he should take two with him, in addition to the one backed with brown paper to show the results of fracture. To ensure a professional appearance, the Unit will make him a 'cover' slide, showing his name and department, and the University crest against a background of the University colour. The crest might become a kind of logo, reappearing in the corner of each slide. The graph showing the statistical analysis can easily be made into a slide, and there are pictures available of the glass in racks on the roof.

Andrew happens to mention that one or two of the audience will not have a technical background, and the Unit staff suggest that he should therefore have a strong visual message about the effect of weathering. Why not use a slide of a cracked window to make the point? Considerable damage had been caused to some disused warehouses in Abimouth by an unusually severe storm during the previous winter, and one of the staff is sure that he could take a series of dramatic photographs for Andrew to choose from. Andrew likes the idea, especially as he wants to impress the Managing Director with the importance of his research.

Back in the Department, Andrew starts to plan his presentation. He considers the two sets of objectives, his and the audience's. He has little doubt about his own: he wants to impress the company with his personal expertise and the good work carried out in the Department, and indeed at the University. He also wants a further contract to be agreed. He has enjoyed carrying out this research, and he knows that he must show that, too.

The company's objectives are less clear. They will naturally want to see value for money, and Andrew is sure that he can demonstrate this, not least in the confidence with which he puts forward his information. He thinks back to meetings he had with company personnel before the research started. They were concerned for their design codes, which might need to be modified if loss of strength makes the glass substantially weaker; there is an obvious anxiety about possible insurance claims if the glass breaks, especially if it falls out and there is personal injury. Andrew suspects that this is the primary concern of the Managing Director, and he feels that it would be wise to make a reference to the problem early in his talk.

The first section is taking shape. After his cover slide, Andrew will start the presentation by showing the slide of the broken window and explaining why there is a problem; he will talk through the precise objective of his research— the need to establish the loss of strength through weathering of 3 mm domestic glass, in the short term. He will refer to the company's design codes, and the need for them to take account of the possibility of the glass losing strength sufficiently to break, with implications for insurance claims.

This seems to be a strong beginning, especially in clarifying the need for the research in the minds of those who have a financial or legal rather than a technical interest. Andrew reckons that it will probably take about five minutes to get this far. He provisionally allocates a further 10 minutes for the technical core of his presentation. He will have to explain how the tests were carried out, showing the slides of the rig and of the glass in the racks. It will be helpful to show specimens of both fresh and weathered glass, and of course he has his disc backed with brown paper, which caused considerable interest at the seminar. He is very much at home with this area of the presentation, but as he plans what he will say, he is aware of how careful he must be with technical terms.

He has two linked problems. If he uses too many technical words and phrases, he will lose the Managing Director; if he resorts to non-technical explanations, he is likely to lose at least one of the French visitors, and perhaps bore the Technical Director. In the end, he decides that he should omit any explanation of uniaxial stress equivalent, and concentrate on the visual impact of the slides and the specimens. He is more worried about the statistical analysis, as this may well be of interest to several members of the audience, but too complicated for others. A handout seems to be the answer: he will use the graph during the presentation, to show the trend, and will prepare a handout of the tables, to be given at the end to anyone who is interested. As long as he remembers to say during the talk that the detailed figures will be available in this way, everyone should be satisfied.

The conclusion of the presentation is more difficult. Andrew needs a strong ending, and very much wants to put in a plea for further funding, but he must put it in the context of the needs of the company. He decides to concentrate on the possible effect of his conclusions on the design codes, to stress that he has looked at the problem in the short term only, and to add that he would be happy to carry the work further. This seems right but a little bit weak. Andrew feels that he needs some visual support at the end of the presentation, but, apart from going back to the 'cover' slide, nothing seems appropriate.

This unease is still at the back of his mind as he begins to prepare his notes. He uses small file cards, on which he writes the key words and phrases of his presentation. To a certain extent, he will use his slides as prompts, but he feels that it would be wise to have some notes as well, particularly as he can include reminders about when to use the slides—and when to give out the printed material. It would be all too easy to forget this at the end of the presentation.

The slides are now ready, and Andrew tries them out in a rehearsal by himself in a lecture room. The quality is good, and he particularly likes the effect of the broken window slide, which seems to pinpoint the main reason for the research. The timing is quite good for a first run through, and Andrew knows that he can refine it over the next few days.

There is still the problem of the end. He needs the equivalent of the broken window slide, to illustrate his results. At home in the evening, he is still worrying about this when he switches on his television and sees a few minutes of a programme about hay fever—a subject near to his heart, as he is affected by pollen each spring. The presenter is talking about the grains of pollen, and shows some under a microscope, and Andrew has an inspiration. A slide of fresh glass under the microscope would be almost invisible, but a slide of weathered glass would show thousands of tiny abrasions—the whole surface would appear to be pitted. The contrast would be clear and striking, and would emphasize the effect of weathering far more strongly than any words.

Andrew has the strong ending that he was looking for, and the Audio-Visual Unit is equally impressed; they will prepare the slide for him. In the meantime, he must think about the questions he is likely to be asked—a stage of preparation which he starts by himself but continues with the help of a colleague. A friend in the same Department, who missed the seminar, is happy to come to a rehearsal; he has a double brief—to make sure that Andrew is presenting his information as effectively as possible, and to ask appropriate questions afterwards. He enjoys what he hears and sees, but reminds Andrew of the need to pause during the presentation, especially when he moves from one aspect of his topic to another. As he rightly says, the Managing Director will need time to assimilate as many of the ideas as possible, and a short breathing space will be helpful. Andrew is glad of the comment, and as he gains confidence he finds it easier to allow silence at strategic moments of his talk.

Together, they consider the questions that Andrew might be asked. One which seems particularly likely is about the weathering of the glass, and how far it is typical of weathering in service. Andrew is not sure how to handle this, as it could potentially undermine the credibility of his investigation; his colleague asks about the precautions which were taken when the glass was installed in the racks, and suggests that Andrew should concentrate on the practical issues: for instance, the racks had deliberately been raised well above ground level, in order to avoid accidental abrasion caused by grit splashing up in heavy rain. This is a good example to give, as it should reassure a questioner about Andrew's awareness of the potential problem. Andrew is glad that he had involved his colleague in his preparations, and grateful that the question has been raised at the rehearsal stage.

The final slide is ready a day or two later, and Andrew is able to link it to his message about future research:

> This is what weathered glass looks like under a microscope, and as you can see, it looks dramatically different from new glass. There's no evidence so far that these effects—and the loss of strength that they suggest—will stop at this. They may get worse. We need to know more about the long-term effects of weathering, the possible continued loss of strength in our glass, and the safety implications of this in the future.

Andrew's presentation is just about ready, and he has timed it at about 17 minutes, giving himself a three-minute margin in case he speaks more slowly in front of the audience. With the help of his colleague, he has predicted the likely questions, and he has good visual support. He feels confident, though a little nervous, and looks professional as he arrives early at the company's premises—hoping that he can have a quick look at the room and the equipment before the audience arrives.

The Technical Director greets him warmly, and Andrew is happy to see a familiar face. The room is appropriately set out, and the equipment, when Andrew checks it, is in good order and easy to use. The blinds are already drawn, and there is a switch to dim the lights before the slides are shown. By the time the introductions are over and the audience has settled to listen, Andrew is beginning to enjoy the experience. All his careful preparation is paying off. He looks at the audience, smiles briefly in greeting them, and introduces himself and his subject.

The presentation itself lasts for 18 minutes, and at the end Andrew asks if his audience has any questions. He is slightly amused that the very first one is the question suggested by his colleague, and his description of his precaution in placing the racks above the level of rain water splashing is accepted without demur. A few minutes later, however, he is faced with a question which he had not foreseen: the Managing Director, who has been following the proceedings with apparent interest, suddenly asks what the financial benefits of the research will be. Andrew says honestly that he does not know, but that the benefits will be seen when more work has been carried out in updating the design codes. His answer is accepted, and he has a good opportunity to stress the need for more research in the future when he is asked by one of the French visitors to estimate the trend in loss of strength beyond the first year's weathering.

Andrew goes quickly back to the slide showing the graph: a visual aid will help to overcome any language barrier. He points to the loss of strength so far, and adds that such a process is likely to be rapid at first but to slow down over time; exact figures could not be given without further research, but this would be the expected pattern.

Before the audience can ask more questions, the Managing Director stands up and closes the meeting. He thanks Andrew for his work, and, to Andrew's particular pleasure, for making a complex technical subject so interesting. He has learnt a lot, and hopes that Andrew's collaboration with the company will continue.

This is, of course, just what Andrew is hoping to hear; the congratulations and thanks of the technical staff present also reassure him that he has managed to put his message across to both specialists and non-specialists in the audience.

Some time later, his hard work on both report and presentation is rewarded when he is asked to continue the research project, investigating the glass and its loss of strength in the longer term.

Key stages

Previous chapters have concluded with Key points; in this case we look back at the key stages which Andrew Poynter has identified in preparing his report and his presentation.

The report

- Andrew identifies the terms of reference, readership and objectives
- The structure of the report is planned, with the aid of a spider diagram; a Contents List is drafted
- The tests are carried out and the results recorded
- The spider diagram is expanded to take in the practical work and its outcome
- The Introduction is drafted
- Sections dealing with the apparatus and the method are prepared
- The statistical analysis and the results are shown in diagrammatic form, with appropriate written explanation
- Conclusions and Recommendations are considered and drafted
- Revision suggests the addition of extra material under Method; this is added
- The Summary is drafted, revised and shortened
- The whole text is thoroughly checked, by Andrew and by a colleague
- A title page is added, the report is printed in its final version, bound and sent to the company

Seminar presentation

- Andrew decides to present his material to his colleagues at an informal seminar

- He decides to use the overhead projector, and to give a demonstration of the work
- He analyses exactly what he wants to gain from the seminar, and the expectations of the audience
- The specimens are prepared, including the glass disc with the paper backing
- Andrew decides what he can sensibly include in a 10-minute talk, and tries it out
- Questions are considered and answers prepared
- Andrew gives his seminar, and later considers what he has learnt from it

Presentation to clients

- Andrew finds out about the audience, the venue and the equipment available
- Appropriate visual aids are discussed; they will be prepared by experts
- Andrew analyses more closely his own and the audience's objectives
- The time is carefully allocated, and a strong beginning prepared
- The decision is taken to use a handout for the technical detail
- The timing is checked in a first run through
- A strong ending is prepared
- A colleague is asked to attend a rehearsal, to check the presentation style and to ask questions
- The timing is now correct, and Andrew feels confident as well as a little nervous
- He arrives early and checks the room and the equipment
- The presentation is given; Andrew is able to handle the questions, and remains aware of the audience and its particular needs
- Both speaker and audience are pleased with the presentation and its reception

Bibliography

Day, Robert A. (1989) *How to write and publish a scientific paper*, Cambridge University Press.

van Emden, Joan (1990) *A Handbook of Writing for Engineers,* Macmillan.

van Emden, Joan and Easteal, Jennifer (1993) *Report Writing*, Stanley Thornes, 2nd edn.

van Emden, Joan and Easteal, Jennifer (1993) *Technical Report Writing*, IEE Professional Brief, 3rd edn, IEE.

Fisher, Barry (1995) *How to document ISO 9000 Quality Systems*, Ramsbury Books.

Goodlad Sinclair (1990) *Speaking Technically*, published privately by the author.

Haslam, Jeremy (1988) *Writing Engineering Specifications*, Spon.

Hennessy, Brendan (1989) *Writing Feature Articles,* Heinemann.

Kirkman, John (1992) *Good Style: writing for science and technology*, Spon.

Kirkman, John (1994) *Guidelines for giving Effective Presentations*, Ramsbury Books.

McRobb, Max, (1989) *Writing Quality Manuals*, IFS publications UK.

The Oxford Dictionary for Writers and Editors (1990) Oxford University Press.

Wieringa, Douglas, Moore, Christopher and Barnes, Valerie (1993) *Procedure Writing*, IEEE Press.

Writers' and Artists' Yearbook (1995), A & C Black.

Index

Visual aids, 116–121
  accuracy of, 121
  appropriate use of, 5–6, 117–119
  quality of, 117
  selection of, 115, 119–121
Visual aids, specific:
  demonstrations, 120
  display material, 118
  flipcharts, 120
  handouts, 121
  LCD panels, 121

overhead projectors, 119
slides, 119–120
video, 120
Voice, use of, 93–99
Volume in speaking, 94–95

Words:
  choice of, 17–20, 68
  meaning and implication of,
    19
  precision in use of, 18